REGIONAL CONFERENCE SERIES IN APPLIED MATHEMATICS

A series of lectures on topics of current research interest in applied mathematics under the direction of the Conference Board of the Mathematical Sciences, supported by the National Science Foundation and published by SIAM.

GARRETT BIRKHOFF, *The Numerical Solution of Elliptic Equations*

D. V. LINDLEY, *Bayesian Statistics — A Review*

Titles in Preparation

R. S. VARGA, *Functional Analysis and Approximation Theory in Numerical Analysis*

The NUMERICAL SOLUTION of ELLIPTIC EQUATIONS

GARRETT BIRKHOFF
Harvard University

SOCIETY for INDUSTRIAL and APPLIED MATHEMATICS
PHILADELPHIA, PENNSYLVANIA 19103

Copyright 1972 by
Society for Industrial and Applied Mathematics
All rights reserved

Printed for the Society for Industrial and Applied Mathematics by
J. W. Arrowsmith Ltd., Bristol 3, England

THE NUMERICAL SOLUTION OF ELLIPTIC EQUATIONS

Contents

Acknowledgments	vii
Preface	ix
Lecture 1 TYPICAL ELLIPTIC PROBLEMS	1
Lecture 2 CLASSICAL ANALYSIS	11
Lecture 3 DIFFERENCE APPROXIMATIONS	19
Lecture 4 RELAXATION METHODS	29
Lecture 5 SEMI-ITERATIVE METHODS	39
Lecture 6 INTEGRAL EQUATION METHODS	47
Lecture 7 APPROXIMATION OF SMOOTH FUNCTIONS	55
Lecture 8 VARIATIONAL METHODS	63
Lecture 9 APPLICATIONS TO BOUNDARY VALUE PROBLEMS	73

Acknowledgments

The material published here grew out of notes for a series of lectures given in a Regional Conference at the University of Missouri at Rolla, supported by the National Science Foundation under a grant to the Conference Board of the Mathematical Sciences. It has been reworked with support from the Office of Naval Research, and has benefited greatly from suggestions by George Fix, Louis Hageman, Robert E. Lynch, John Rice, Richard Varga, Gene Golub, James Bramble, Dieter Gaier, and others.

GARRETT BIRKHOFF

Preface

These lecture notes are intended to survey concisely the current state of knowledge about solving elliptic boundary-value and eigenvalue problems with the help of a modern computer. For more detailed accounts of most of the relevant ideas and techniques, the reader is referred to the general references listed following this preface, to the basic references listed at the end of each lecture, and to the many research papers cited in the footnotes.

To some extent, these notes also provide a case study in *scientific computing*, by which I mean the art of utilizing physical intuition, mathematical theorems and algorithms, and modern computer technology to construct and explore realistic models of (perhaps elliptic) problems arising in the natural sciences and engineering.

As everyone knows, high-speed computers have enormously extended the range of effectively solvable partial differential equations (DE's). However, one must beware of the myth that computers have made other kinds of mathematical and scientific thinking obsolete. The kind of thinking to be avoided was charmingly expressed by C. B. Tompkins twenty years ago in his preface to [1], as follows:

"I asked Dr. Bernstein to collect this set of existence theorems during the parlous times just after the war when it was apparent to a large and vociferous set of engineers that the electronic digital calculating machines they were then developing would replace all mathematicians and, indeed, all thinkers except the engineers building more machines.

"Many of the problems presented were problems involving partial differential equations. The solution, in many cases, was to be brought about (according to the vociferous engineers) by:

(1) buying a machine;
(2) replacing the differential equation by a similar difference equation with a fine but otherwise arbitrary grid;
(3) closing the eyes, mumbling something about truncation error and round-off error; and
(4) pressing the button to start the machine."

The myth so wittily ridiculed by Tompkins contains a grain of truth, nevertheless. One *can* approximate almost any DE by a difference equation (ΔE) with an arbitrarily high "order of accuracy." In particular, one can approximate any *linear* DE by an approximating system of *simultaneous linear equations* which can, in principle, be solved algebraically.

In the 1950s, many techniques were invented for solving elliptic problems approximately in this way; an excellent discussion of them is contained in Forsythe and Wasow [FW, Chaps. 20–25].[1] The first five lectures below cover roughly the same material, but in more condensed style and with up-dated references. I shall devote my second lecture to a brief survey of a few facts about classical analysis which relate most specifically to elliptic DE's. My third lecture will be largely concerned with what classical analysis can say about the accuracy of difference approximations. My next two lectures will be devoted to recent developments in numerical algebra, and especially to the SOR and ADI iterative techniques for solving elliptic DE's.[2] Since 1965, the emphasis has shifted to variational methods and techniques of piecewise polynomial approximation (by "finite elements", "splines," etc.), for solving elliptic problems. My next three lectures will be primarily concerned with these methods. Note that both approximation theory and "classical" real and complex numerical algebra play essential roles in scientific computing; so does the "norm" concept of functional analysis.

However, I shall say little about classical algebra or (modern) functional analysis, because an adequate discussion of the first would lead too far afield, and because Professor Varga will cover the second in his lectures. Neither shall I say much about organization of computers, even though designers of large and frequently used "production" codes must take this into account.

The sixth lecture, which builds on the ideas introduced in the second lecture, deals with the adaptation to computers of deeper techniques from classical analysis. For these methods, which tend to apply only to special classes of problems, the book by Kantorovich and Krylov [KK] is the best reference.

The next two lectures, Lectures 7 and 8, center around recent applications of piecewise polynomial ("finite element," "Hermite," or "spline") approximations to the solution of elliptic problems having variational formulations. The last lecture reviews briefly the current status of a number of specific classes of problems, in the light of the material presented in Lectures 1–8.

Throughout, results from the following list of general references will be utilized freely. More special lists of references will also be given at the end of each chapter.

GARRETT BIRKHOFF

---

[1] Capital letters in square brackets designate general references listed after this preface.

[2] However, I shall not discuss general techniques for computing algebraic eigenvectors and eigenvalues, because they are so masterfully discussed by Wachspress [W] and in R. S. Martin and J. H. Wilkinson, Numer. Math., 11 (1968), pp. 99–110 and G. Peters and J. H. Wilkinson, SIAM J. Numer. Anal., 7 (1970), pp. 479–492.

General References

[Az] A. K. AZIZ, editor, *Lecture Series in Differential Equations*, vol. II, Van Nostrand Mathematical Study no. 19, Princeton, New Jersey, 1969.
[BV] G. BIRKHOFF AND R. S. VARGA, editors, *Numerical Solution of Field Problems in Continuum Physics*, SIAM–AMS Proceedings II, American Mathematical Society, Providence, 1969.
[C]   L. COLLATZ, *Numerical Treatment of Differential Equations*, 3rd ed., Springer, Berlin, 1960.
[CH] R. COURANT AND D. HILBERT, *Methods of Mathematical Physics*, vols. I, II, Interscience, New York, 1953, 1962.
[F]   L. FOX, *Numerical Solution of Ordinary and Partial Differential Equations*, Addison-Wesley, Reading, Massachusetts, 1962.
[FW] G. E. FORSYTHE AND W. R. WASOW, *Finite Difference Methods for Partial Differential Equations*, John Wiley, New York, 1960.
[K]   O. D. KELLOGG, *Potential Theory*, Springer, Berlin, 1929.
[KK] L. V. KANTOROVICH AND V. I. KRYLOV, *Approximate Methods of Higher Analysis*, Noordhoff–Interscience, New York–Groningen, 1958.
[V]   R. S. VARGA, *Matrix Iterative Analysis*, Prentice-Hall, Englewood Cliffs, New Jersey, 1962.
[V'] ———, *Functional Analysis and Approximation Theory in Numerical Analysis*, Regional Conference Series in Applied Math. 3, SIAM Publications, Philadelphia, 1971.
[W]   E. WACHSPRESS, *Iterative Solutions of Elliptic Systems*, Prentice-Hall, Englewood Cliffs, New Jersey, 1966.
[Wi] J. H. WILKINSON, *The Algebraic Eigenvalue Problem*, Oxford University Press, Oxford, 1965.

LECTURE 1

Typical Elliptic Problems

1. Two-endpoint problems. My aim in these lectures will be to describe a variety of powerful and sophisticated numerical techniques which have been used to obtain approximate solutions of elliptic differential equations and systems of equations on high-speed computers. My first lecture will be devoted to describing some typical physical problems to which these methods apply. I do this because the effective use of computers often requires physical intuition to help one decide how to formulate physical problems, which parameters are most important over which range, and whether an erratic computer output is due to physical or to numerical instability. For this reason, I shall devote my first lecture to the intuitive physical background of some of the most commonly studied elliptic problems of mathematical physics.

In elliptic problems, one is given a partial differential equation (partial DE) to be satisfied in the interior of a region R, on whose boundary ∂R additional boundary conditions are also to be satisfied by the solution. In the one-dimensional analogue of ordinary DE's, the region is an interval whose boundary consists of two endpoints. Therefore, two-endpoint problems for ordinary DE's may be regarded as boundary value problems of "elliptic" type. (By contrast, well-set initial value problems for partial DE's are typically of parabolic or hyperbolic type.)

The simplest two-endpoint problem concerns a transversely *loaded string*, in the small deflection or *linear approximation* (cf. § 8). If the string (assumed horizontal) is under a constant tension T, then the vertical deflection y induced by a load exerting a transverse force $f(x)$ per unit length satisfies the ordinary DE

$$(1) \qquad -y'' = f(x)/T \qquad \text{(force in the } y\text{-direction)}.$$

If the endpoints are fixed, then the deflection satisfies also the two endpoint conditions

$$(1') \qquad y(0) = y(c) = 0.$$

The formal solution of the system (1)–(1') is elementary: One must first find an antiderivative $g(x)$ of $f(x)$; then an antiderivative $h(x)$ of $g(x)$. Both of these are easily computed by numerical quadrature (e.g., by Simpson's rule). The general solution of (1) is then $h = h(x) + ax + b$; the boundary conditions (1') are satisfied by some unique choice of a and b, giving the solution.

The problem of a longitudinally *loaded spring* is similar. If $p(x)$ is the stiffness of the spring, and $f(x)$ is the load per unit length, then the appropriate DE for the longitudinal deflection $y(x)$ is

(2) $$-[p(x)y']' = f(x), \qquad p(x) > 0,$$

and one can again impose the fixed endpoint conditions (1') or, more generally,

(2') $$y(0) = y_0, \quad y(c) = y_1.$$

As a third example, we consider *Sturm-Liouville systems*. These typically arise from separating out the time variable from simply harmonic solutions of time-dependent problems such as that of a vibrating string. They are defined by *homogeneous* linear DE's of the form

(3) $$[p(x)y']' + (\lambda p(x) + q(x))y = 0,$$

in which λ is a parameter, and homogeneous linear boundary conditions of the form (1') or, more generally,

(3') $$\alpha_0 y(0) + \beta_0 y'(0) = \alpha_1 y(c) + \beta_1 y'(c) = 0.$$

It is well known that any S-L system admits an infinite sequence $\{\lambda_i\}$ of real *eigenvalues* $\lambda_1 < \lambda_2 < \lambda_3 < \cdots$ with $\lambda_n \to \infty$, for which there are nontrivial solutions called *eigenfunctions*.

In summary, we have described above two *boundary value* problems and one *eigenvalue–eigenfunction* problem which have important higher-dimensional elliptic analogues.

2. Dirichlet and related problems. The most deeply studied elliptic boundary value (B.V.) problem is the Dirichlet problem, which can be described in physical terms as follows.

Let a homogeneous solid occupy a bounded region R in n-dimensional space, and let its boundary ∂R be kept at a specified temperature distribution $g(\mathbf{y})$ ($\mathbf{y} \in \partial R$). What will be the equilibrium temperature distribution $u(\mathbf{x})$ in the interior? Under the physically plausible (and fairly realistic) assumption that the flow ("flux") of heat at any point is proportional to the temperature gradient ∇u there, one can show that the temperature must satisfy the *Laplace equation*:

(4) $$\nabla^2 u = \sum_{i=1}^{n} \frac{\partial^2 u}{\partial x_i^2} = 0 \quad \text{in} \quad R \qquad (n \text{ space dimensions}).$$

The *Dirichlet problem* is to find a function which satisfies (4) in R and

(4a) $$u(\mathbf{y}) = g(\mathbf{y}) \quad \text{on} \quad \partial R,$$

and is continuous in the closed domain $R \cup \partial R$.

The Laplace equation (4) arises in a variety of other physical contexts, often in combination with other boundary conditions. In general, a function which satisfies (4) is called *harmonic* (in R); and the study of harmonic functions (about which I shall say more in the next lecture) is called *potential theory*.

For example (see [K] or [7][1]), the Laplace equation (4) is satisfied in empty regions of space by gravitational, electrostatic and magnetostatic potentials. Thus the electrostatic potential due to a charged conductor satisfies (4) in the *exterior* of R and, in suitable units,

(4b) $$u = 1 \quad \text{on} \quad \partial R, \quad u \sim C/r \quad \text{as} \quad r \to \infty.$$

The problem of solving (4) and (4b) is called the *conductor* problem, and the constant C is called the *capacity* of the conductor. Many other problems of potential theory are described in Bergman–Schiffer [1].

Likewise, the irrotational flows of an incompressible fluid studied in classical hydrodynamics [5, Chaps IV–VI] have a "velocity potential" which satisfies (4). For liquids of (nearly) constant density, this remains true under the action of gravity, a fact which makes (4) applicable also to some problems of petroleum reservoir mechanics in a homogeneous medium (soil).[2]

However, the boundary conditions which are appropriate for these applications are often quite different from those of (4a). Thus, in hydrodynamical applications, the usual boundary conditions amount to specifying *normal derivatives*,[3] or

(4c) $$\partial u/\partial n = h(\mathbf{y}) \quad \text{on} \quad \partial R.$$

The problem of finding a harmonic function with given normal derivatives on the boundary is called the *Neumann problem*.

More generally, in the theory of heat conduction, it is often assumed that a solid loses heat to the surrounding air at a rate roughly proportional to the excess surface temperature (Newton's "Law of Cooling"). This leads one to try to solve (4) for the boundary conditions

(4d) $$\partial u/\partial n + ku = kg(\mathbf{y}) = h(\mathbf{y}) \quad \text{on} \quad \partial R.$$

(If the conductor is cut out of sheet metal, one may look instead for functions which satisfy (4a) and the modified Helmholtz DE: $u_{xx} + u_{yy} = ku, k > 0$, instead of (4) inside the conductor.)

3. Membranes; source problems. Potential theory is concerned not only with harmonic functions, but also with solving the Poisson equation

(5) $$-\nabla^2 u = f(\mathbf{x})$$

in free space and in bounded domains subject to various boundary conditions. Evidently, the Poisson equation (5) is the natural generalization to $n > 1$ space dimensions of the DE (1) for the loaded string problem. Indeed, when $n = 2$, the DE (5) is satisfied by the transverse deflection $z = u(x, y)$ of a horizontal *membrane*

---

[1] Numbers in square brackets refer to the references listed at the end of the lecture; letters in square brackets to the list of general references after the preface.

[2] See [8]; also P. Ya. Polubarinova-Kochina, *Advances in Applied Mathematics*, 2 (1951), pp. 153–221, and A. E. Scheidegger, *Physics of Flow through Porous Media*, Macmillan, 1957.

[3] Here and below, $\partial/\partial n$ means *exterior* normal derivative.

(or "drumhead") under uniform lateral tension T, which supports a load of $Tf(\mathbf{x})$ per unit area.

For such a membrane, held in a fixed rigid frame, the appropriate boundary condition is

(5') $\qquad u = 0 \quad \text{on} \quad \partial R, \quad \text{the membrane boundary}.$

To solve (5) in R subject to the boundary condition (5') will be one of our main concerns below.

When $n = 3$, the DE (5) is also satisfied by the gravitational potential of a continuous distribution of matter with density $4\pi f(\mathbf{x})$ per unit volume. Likewise, it is satisfied by the electrostatic potential of a continuous charge distribution having this density. These observations lead to other boundary value problems in the Poisson equation.

A more general elliptic DE is

(6) $\qquad\qquad -\nabla \cdot [p(\mathbf{x})\nabla u] = f(\mathbf{x}).$

It has the notable property of being *self-adjoint*, which implies that its Green's function $G(\mathbf{x}, \boldsymbol{\xi})$ (see Lecture 2, § 4) is *symmetric* in the sense that $G(\mathbf{x}, \boldsymbol{\xi}) = G(\boldsymbol{\xi}, \mathbf{x})$, and that its *eigenvalues* are *real*.

This DE is satisfied by the temperature distribution $u(\mathbf{x})$ in a solid having space-dependent thermal conductivity $p(\mathbf{x})$, in which heat is being produced at the rate of $f(\mathbf{x})/4\pi$ per unit volume and time. Since one may think of $f(\mathbf{x})$ as representing a *source* of heat, the DE (6) for suitable boundary conditions is often said to define a *source problem*. Such source problems arise typically in the analysis of diffusion phenomena. The DE (6) also arises as Darcy's Law in petroleum reservoir mechanics [8, p. 242], with p the (soil) permeability, u the pressure, and $f(\mathbf{x}) = \rho g$ constant. In practice, p may vary by orders of magnitude—like thermal and electrical conductivity.

Diffusion with convection. Another important family of elliptic DE's describes convection with diffusion. For any velocity-field $(U(x, y), V(x, y))$ with divergence $U_x + V_y = 0$, the DE

$$U\varphi_x + V\varphi_y = \alpha \nabla^2 \varphi, \qquad\qquad \alpha > 0,$$

can be interpreted in this way. One should remember that, although this DE is elliptic, convection dominates diffusion in the long run, so that in many respects its solutions behave like solutions of the hyperbolic DE $U\varphi_x + V\varphi_y = 0$, but smoothed or "mollified" locally.

4. Reduced wave equation. The equation of a transversely *vibrating membrane* is

$$z_{tt} = c^2(z_{xx} + z_{yy}),$$

where $c = (T/\rho)^{1/2}$ is the wave velocity (T the tension and ρ the density per unit area of the membrane, both assumed constant). Simple harmonic (in time) oscillations of such a membrane are clearly given by setting

$$z(x, y, t) = u(x, y) \begin{Bmatrix} \sin \\ \cos \end{Bmatrix} kt,$$

where $u(x, y)$ is a solution of the Helmholtz or *reduced wave equation*

(7) $$\nabla^2 u + k^2 u = 0, \qquad \nabla^2 = \sum_{i=1}^{n} \frac{\partial^2 u}{\partial x_i^2},$$

with $n = 2$ and $x_1 = x$, $y_1 = y$. Hence, to find the possible simply harmonic vibrations of a membrane held in a rigid frame having a given contour ∂R, we must find the solutions of the Helmholtz equation (7) subject to the boundary condition

(7a) $$u = 0 \quad \text{on} \quad \partial R.$$

Similarly, in three-dimensional space, let $u(\mathbf{x}) = u(x, y, z)$ be a solution of the reduced wave equation (7) with $n = 3$ in a bounded domain R with boundary ∂R, and let

(7b) $$\partial u/\partial n = 0 \quad \text{on} \quad \partial R.$$

Then $p(x, y, z, t) = u(x, y, z) \begin{Bmatrix} \sin \\ \cos \end{Bmatrix} kt$ describes the pressure variations (from ambient pressure) in a standing sound wave with frequency $ck/2\pi$ in a room (or organ pipe) having the specified (rigid) boundary ∂R.

Just as in the case of Sturm-Liouville systems (see § 1), each of the systems (7)–(7a) and (7)–(7b) has a sequence of nontrivial solutions called the eigenfunctions of the system, whose eigenvalues $\lambda_j = k_j^2$ are positive (or zero) and can be arranged in ascending order:

$$\lambda_1 \leq \lambda_2 \leq \lambda_3 \leq \cdots, \quad \lambda_n \uparrow \infty.$$

I shall discuss this "Ohm-Rayleigh principle" in the next lecture; various classical examples are worked out in textbooks on sound.[4]

Maxwell's equations. By separating out the spatial variation of simply harmonic (in time) "standing wave" solutions of Maxwell's equations for electromagnetic waves in a homogeneous medium, one is led to other solutions of the reduced wave equation. However, quantitative results about wave guides and scattering are still usually obtained by analytical methods.[5] High-speed digital computers are just beginning to be useful for solving Maxwell's equations (cf. Lecture 9).

---

[4] P. M. Morse, *Vibrations and Sound*, McGraw-Hill, New York, 1936.

[5] See R. E. Collin, *Field Theory of Guided Waves*, McGraw-Hill, New York, 1960, Chap. 8; L. Lewin, *Advanced Theory of Wave Guides*, Iliffe, London, 1951; N. Markuvitz, *Waveguide Handbook*, McGraw-Hill, New York, 1951.

5. Thin beams; splines.

The problems discussed so far have all involved *second order* elliptic DE's. In solid mechanics, *fourth order* elliptic DE's and systems are more prevalent.

The simplest such problems refer to the small deflections of a *thin beam* or "rod" by an applied transverse "load" or force distribution. This problem was solved mathematically by the Bernoullis and Euler, who assumed that the beam or "elastica" was homogeneous, i.e., had the same physical characteristics in all cross-sections. From Hooke's law, D. Bernoulli deduced in 1706 that, in the linear or "small deflection" approximation (see § 8), the deflection of the centerline of the beam should satisfy (see [9]):

$$(8) \qquad u^{\text{iv}}(x) = d^4 u/dx^4 = f(x), \qquad a \leq x \leq b.$$

Here $f(x)$ is the quotient of the applied transverse load per unit length by the "stiffness" of the beam, whose undeflected centerline is supposed to extend along the x-axis from $x = a$ to $x = b$.

Thin beam problems can involve various sets of endpoint conditions, notably the following [CH, pp. 295–296]:

$$(8a) \qquad u''(a) = u'''(a) = u''(b) = u'''(b) = 0 \qquad \text{(free ends)}$$

$$(8b) \qquad u(a) = u''(a) = u(b) = u''(b) = 0 \qquad \text{(simply supported ends)}$$

$$(8c) \qquad u(a) = u'(a) = u(b) = u'(b) = 0 \qquad \text{(clamped ends)}.$$

Regardless of the endpoint condition selected, the general solution of the DE (8) is the sum of any particular solution and some *cubic polynomial*, since the general solution of the ordinary DE $u^{\text{iv}}(x) = 0$ is a cubic polynomial. Hence, to solve any of the above two-endpoint problems for a thin beam, one can proceed as in solving (1).

Namely, one can first compute a particular solution $U(x)$ of (8) by performing four successive quadratures on $f(x)$ numerically (e.g., by Simpson's rule). One then forms

$$(9) \qquad u(x) = U(x) + c_0 + c_1 x + c_2 x^2 + c_3 x^3,$$

regarding the coefficients c_j as unknown coefficients to be determined from the four endpoint conditions.

Cubic splines. A very useful special case corresponds to "point-loads" concentrated at some sequence of points x_i:

$$\pi: a = x_0 < x_1 < x_2 < \cdots < x_{r-1} < x_r = b \quad \text{of} \quad [a, b].$$

Since, for any $c < d$,

$$u'''(d) - u'''(c) = \int_c^d u^{\text{iv}}(x) = \int_c^d f(x)\, dx,$$

a total load of w_i concentrated at x_i may be expected to produce a jump of $w_i = u'''(x_i^+) - u'''(x_i^-)$ in the third derivative of the deflection function, whose second derivative is presumably continuous. This suggests that the solutions are given by the following class of functions.

DEFINITION. A cubic spline function on $[a, b]$ with *joints* (or "knots") at the x_i, $i = 1, \cdots, r - 1$, is a function $u \in C^2[a, b]$ which is expressible on each segment (x_{i-1}, x_i) by a cubic polynomial $p_i(x) = \sum_{k=0}^{3} a_{ik} x^k$.

Splines have been used by naval architects for many years to generate mechanically smooth curves which pass through (or "interpolate" to) preassigned points; as we shall see in Lectures 7 and 8, "spline functions" are also useful for computing accurate numerical solutions of elliptic DE's.

6. Plates and shells. Solid mechanics provides many challenging elliptic problems for the mathematician to solve. One of the simplest of these is provided by Kirchhoff's theory of a transversely loaded flat plate. The transverse deflection satisfies the deceptively simple-looking biharmonic equation

$$(10) \qquad \nabla^4 u = f(x, y).$$

As in the one-dimensional analogue of the thin beam, one may have any of a fairly large variety of boundary conditions.

It may surprise readers to know that the DE(10) is the Euler-Lagrange DE associated with the variational condition $\delta J = 0$ for a whole family of integrals J: (10) is implied by

$$(11) \qquad \delta \left[\int \int \{ (\nabla^2 u)^2 + (1 - \nu)[u_{xx} u_{yy} - u_{xy}^2] \} \, dx \, dy \right] = 0$$

for *any* value of the "Poisson ratio" ν (see [11]).

A parallel-loaded homogeneous plate with body force "load" potential $V(x, y)$ has stress components σ_x, σ_y and τ_{xy} most easily expressed in terms of the Airy stress function $\varphi(x, y)$:

$$\sigma_x = \varphi_{yy} + V, \quad \sigma_y = \varphi_{xx} + V, \quad \tau_{xy} = -\varphi_{xy}.$$

The conditions for static equilibrium are given by the compatibility relations

$$\nabla^4 \varphi + \frac{1 - 2\nu}{1 - \nu} \nabla^2 V = 0.$$

If V is harmonic, then φ is biharmonic.

Curvilinear elastic shells satisfy much more complicated but analogous systems of linear elliptic equations with variable coefficients.

7. Multigroup diffusion. Another important area of application for numerical methods is provided by the steady state multigroup diffusion equations of nuclear reactor theory. These constitute a *cyclic* system of DE's for source problems, in an idealized thermal reactor, typically of the form [2]:

$$(12) \qquad \sigma_1^* \varphi_1 - \nabla \cdot ([D(x) \nabla \varphi]) = \nu \sigma_n \varphi_n$$

and

(12') $$\sigma_i^* \varphi_i - \nabla \cdot [D_i(\mathbf{x})\nabla \varphi_i] = \sigma_{i-1}\varphi_{i-1}$$

for $i = 2, \cdots, n$. Here the coefficients σ_i^* and $\sigma_i \leq \sigma_i^*$ are typically piecewise constant. These DE's are to hold in the "reactor domain" R; on the boundary ∂R of R, it is assumed that

(13) $$l_i \partial \varphi_i / \partial n + \varphi_i = 0, \qquad i = 1, \cdots, n, \quad l_i > 0.$$

In the preceding DE's, the dependent variable $\varphi_i(x)$ stands for the "flux" level at $\mathbf{x}$ of neutrons of the ith velocity group and it equals $v_i N_i(\mathbf{x})$, where v_i is the (nominal) mean velocity of neutrons of the ith velocity group and $N_i(\mathbf{x})$ is the expected neutron density (population per unit volume) in the vicinity of $\mathbf{x}$; the D_i are the mean "diffusivities" of neutrons of the ith velocity group; σ_i^* and σ_i are the (macroscopic) absorption and down-scattering cross-sections; v is the mean neutron yield per fission.

The problem is an *eigenfunction* problem; of greatest practical interest are the smallest eigenvalue v_0 (the critical yield per fission) and the associated (positive) *critical flux distribution*.

8. Some nonlinear problems. Many important elliptic problems are *non*linear; I shall here describe only a few examples of such problems.

Probably the simplest nonlinear elliptic problem is that of a loaded string or *cable*. If we use the exact expression for the curvature $\kappa = y''/(1 + y'^2)^{3/2}$, the DE of a loaded string under a horizontal tension T_0 and vertical load $w(x)$ per unit length is

(14) $$T_0 y'' = w(x)(1 + y'^2)^{1/2}.$$

The case of a catenary is $w(x) = (1 + y'^2)^{1/2}$.

Only slightly less simple is the nonlinear thin beam problem, whose DE is (in terms of arc-length s):

(15) $$\frac{d^2\varphi}{ds^2} + R \sin \varphi = 0, \qquad \frac{dy}{dx} = \tan \varphi.$$

Its solutions are described in detail in Love's *Elasticity*, § 262.

Plateau problem. The simplest nonlinear elliptic problem whose solution is a function of two independent variables is probably the Plateau problem [CH, vol. 2, p. 223]. In its simplest form, the problem is to minimize the *area*

(16) $$A = \iint (1 + z_x^2 + z_y^2)^{1/2} \, dx \, dy$$

of a variable surface spanned by a fixed simple closed curve $\gamma: x = x(\theta), y = y(\theta), z = z(\theta)$. Physically, this surface can be realized by a thin *soap-film* spanned by a wire loop tracing out the curve γ (a special-purpose "analogue computer").

The associated Euler-Lagrange variational equation is

(17) $$[1 + z_y^2]z_{xx} - 2z_x z_y z_{xy} + [1 + z_x^2]z_{yy} = 0,$$

which clearly reduces to the Laplace DE for a nearly flat surface, with $z_x \ll 1$, $z_y \ll 1$. This is also the DE of a surface with *mean curvature zero*.

A related nonlinear problem is that of determining (e.g., computing) the surface or surfaces with given *constant mean curvature* $(\kappa_1 + \kappa_2)/2 = M$ spanned by γ.

Nonlinear heat conduction. Actually, conductivity and specific heat are temperature-dependent, while heat transfer rates in fluids depend on the temperature gradient as well as the temperature. Therefore, more exact mathematical descriptions of heat conduction lead to nonlinear DE's.

Of these, $\nabla^2 u + e^u = 0$ has been a favorite among mathematicians because of its simplicity, but it is by no means typical. Some idea of the complexity of real heat transfer problems can be obtained by skimming through [3, Chap. 26].

9. Concluding remarks. Indeed, I want to emphasize the fact that only extremely simple or extremely important scientific and engineering problems can be profitably treated on today's computers. In my lectures, I shall emphasize such very simple problems, because their theory and computational techniques for solving them are relatively far advanced and well correlated with numerical results. In doing this, I shall try to steer a middle course between extremely general "numerical analysis without numbers," in which theorems typically refer to systems of rth order DE's in n independent variables, and "numbers without analysis," alias "experimental arithmetic."

REFERENCES FOR LECTURE 1

[1] S. BERGMAN AND M. SCHIFFER, *Kernel Functions and Elliptic Differential Equations in Mathematical Physics*, Academic Press, New York, 1953.
[2] S. GLASSTONE AND M. EDLUND, *Elements of Nuclear Reactor Theory*, Van Nostrand, Princeton, New Jersey, 1952.
[3] MAX JAKOB, *Heat Transfer*, vols. I, II, John Wiley, New York, 1950, 1957.
[4] JAMES JEANS, *The Mathematical Theory of Electricity and Magnetism*, 5th ed., Cambridge University Press, London, 1941.
[5] H. LAMB, *Hydrodynamics*, 6th ed., Cambridge University Press, London, 1932.
[6] R. E. LANGER, editor, *Frontiers of Numerical Mathematics*, University of Wisconsin Press, Madison, 1960.
[7] P. M. MORSE AND H. FESHBACH, *Methods of Theoretical Physics*, vols. I, II, McGraw-Hill, New York, 1953.
[8] M. MUSKAT, *Flow of Homogeneous Fluids Through Porous Media*, McGraw-Hill, New York, 1937.
[9] J. L. SYNGE AND B. A. GRIFFITH, *Principles of Mechanics*, 2nd ed., McGraw-Hill, New York, 1949.
[10] S. TIMOSHENKO AND J. N. GOODIER, *Theory of Elasticity*, McGraw-Hill, New York, 1951.
[11] S. TIMOSHENKO AND S. WOINOWSKY-KRIEGER, *Theory of Plates and Shells*, McGraw-Hill, New York, 1959.

LECTURE 2

Classical Analysis

1. Classical methods. Nineteenth century mathematical physicists displayed enormous ingenuity in expressing particular solutions of linear partial DE's with constant coefficients in terms of functions of one variable, and of series or integrals involving known special functions of one variable. Their most versatile tool consisted in "separating variables," as in Examples 1–3 to follow.

Though classical methods are especially fruitful for linear DE's with constant coefficients, they also perennially provide significant new solutions of important nonlinear problems.[1] Moreover they play an essential role in scientific computing. Thus numerical methods are generally much less effective than asymptotic and perturbation methods for treating DE's with very small ($\varepsilon \ll 1$) or very large ($\lambda \gg 1$) coefficients. Moreover, many algorithms for solving (elliptic) DE's are inspired by relevant variational principles, series expansions and asymptotic formulas from classical analysis.

Finally, particular solutions of linear partial DE's and ΔE's with constant coefficients play a very special role in contemporary numerical analysis, by providing *model problems* whose exact solutions can be compared with those computed by difference approximations or other general numerical methods. Among such particular solutions, the following are especially noteworthy.

Example 1. Each solution of the Dirichlet problem for $\nabla^2 u = 0$ in the unit disc is given by expanding its (periodic) boundary values in Fourier series:

$$(1) \qquad u(r, \theta) = a_0 + \sum_{k=1}^{\infty} (a_k \cos k\theta + b_k \sin k\theta).$$

The solution is then

$$(1') \qquad u(r, \theta) = a_0 + \sum_{k=1}^{\infty} r^k (a_k \cos k\theta + b_k \sin k\theta).$$

Example 2. The solution of the Poisson equation $-\nabla^2 u = f(x, y)$ in the unit square $S: 0 < x, y < 1$ is easily found by expressing $f(x, y)$ in S as a double sine series:

$$(2) \qquad f(x, y) = \sum_{k=1}^{\infty} \sum_{l=1}^{\infty} c_{kl} (\sin k\pi x \sin l\pi y).$$

---

[1] See, for example, P. Neményi, Advances in Applied Mechanics, vol. 2, Academic Press, New York, 1951, pp. 123–151; R. Berker, Handbuch der Physik, VIII/2, pp. 1–384; W. F. Ames, *Nonlinear Partial Differential Equations in Engineering*, Academic Press, New York, 1965.

The solution is

$$u(x, y) = \frac{1}{\pi^2} \sum_{k=1}^{\infty} \sum_{l=1}^{\infty} \frac{c_{kl}}{(k^2 + l^2)} \sin k\pi x \sin l\pi y. \tag{3}$$

Example 3. Consider the eigenfunction problem for the Helmholtz equation $\nabla^2 u + \lambda u = 0$ in the unit square S, for the homogeneous boundary condition $u \equiv 0$ on ∂S. Its eigenfunctions are $u_{k,l}(x, y) = \sin k\pi x \sin l\pi y$, and its eigenvalues correspondingly $\lambda_{k,l} = (k^2 + l^2)\pi^2$, $k, l = 1, 2, 3, \cdots$.

Many other more sophisticated separations of variables (e.g., into spherical harmonics and Lamé functions [K]) have been developed by mathematical physicists. However, it has gradually become apparent that this tool, though versatile, is not truly general, and that its possibilities have been almost completely exhausted by classical analysts.[2] Indeed, the main advantage of the numerical methods to be described in Lectures 3 to 9 below consists in their relative *generality*. But in any case, I shall devote this lecture to classical analysis and shall return to it in Lectures 6 and 7.

2. Conformal mapping. In § 1, we discussed elliptic problems which can be broken down so as to be tractable using tables of functions of one *real* variable. Other elliptic problems, especially the Dirichlet and similar problems involving the Laplace equation, are amenable to solution in terms of functions of one *complex* variable.

To see this, recall that the real and imaginary parts of any complex analytic function $w = f(z)$ of a *complex* variable z are conjugate harmonic functions of two variables, and conversely. Therefore, potential theory (which is the study of harmonic functions) can be viewed as generalizing some aspects of complex analysis from functions of two variables to functions of n variables. (The study of harmonic functions of one variable is trivial: they are all linear, of the form $u = ax + b$.)

Clearly, the set of harmonic functions $h(x, y)$ is invariant under (direct) one-one *conformal mappings* of the form $t = f(z)$, $z = x + iy$ and f any complex analytic function. Also, since $|du/dt| = |du/dz|/|dt/dz|$, normal derivatives are divided by $|f'(z)|$ under such a conformal mapping. This makes it easy to transform boundary value problems in a region into similar problems in any conformally equivalent region.

Moreover, it is easy to write down formulas which will transform the upper half-plane into a circular disc (machine transformation), a polygon of arbitrary shape (Schwarz–Christoffel transformation), and many other familiar simply connected domains.[3] Indeed, the fundamental theorem of conformal mapping (alias Riemann's mapping theorem) asserts that *every* simply connected compact domain can be so mapped; we shall study the effective computational implementation of this theorem in Lecture 6, § 6 and § 7.

---

[2] See L. P. Eisenhart, Ann. of Math., 35 (1934), pp. 284–305.
[3] See Z. Nehari, *Conformal Mapping*, McGraw-Hill, New York, 1952.

A similar existence theorem holds for other compact Riemannian surfaces (2-spreads); to prove this was the concern of Hilbert's 19th problem and Korn's theorem. Other generalizations apply to pseudo-conformal maps, a subject with many applications.[4]

The method of conformal mapping is much less fruitful in $n > 2$ dimensions, because all conformal transformations of R^n, $n > 2$, are generated by *inversions*, and carry spheres (for example) into spheres or planes. (This is Liouville's theorem.) However, there exist a number of important elliptic problems involving spheres, for which elegant solutions are provided by Kelvin transformations (see [K, pp. 231–2] and [8]).

3. Arithmetic mean and maximum principles. Harmonic functions in any number of dimensions can be characterized as continuous functions which satisfy Gauss' theorem of the arithmetic mean:

Arithmetic mean principle (Gauss–Koebe). A function $u(\mathbf{x})$ continuous in an open region R satisfies $\nabla^2 u = 0$ there if and only if, given a sphere S in R with center $\mathbf{a}$,

$$(4) \qquad u(\mathbf{a}) = \left[\int_S u(\mathbf{x})\, dS \right] \Big/ \Omega(S) \qquad (\Omega(S) = \text{hyperarea of } S).$$

From this in turn one can derive the following principle.[5]

Maximum principle. Let $u(\mathbf{x})$ be harmonic in a compact domain $R = R \cup \partial R$. Then the maximum and minimum values of u are assumed on ∂R.

This maximum principle has been greatly extended and sharpened.[6] It has many important applications. For example, it not only implies the *uniqueness* of solutions for given boundary values; it also implies the continuous dependence of solutions on their boundary values in the uniform norm.

From the converse of the theorem of the arithmetic mean there also follows the Weierstrass convergence theorem: any uniform limit of harmonic functions is harmonic.

4. Green's functions. The concept of the *Green's function* of a linear differential operator L (for specified linear homogeneous boundary conditions) is one of the most suggestive concepts of the theory of partial DE's. Intuitively, $G(\mathbf{x}, \xi)$ is that function which (for fixed ξ) satisfies $L(G) = \delta(\mathbf{x} - \xi)$. It follows (in the sense of the theory of distributions) that the solution of $L[u] = f$ for these same boundary

---

[4] L. Bers, *Mathematical Aspects of Subsonic and Transonic Gas Dynamics*, John Wiley, New York, 1958.

[5] See L. V. Ahlfors, *Complex Analysis*, 2nd ed., McGraw-Hill, New York, 1953, which also contains a very readable proof of Riemann's mapping theorem.

[6] See [9]; D. Gilbarg, Proc. Symposia Pure Math., vol. IV, American Mathematical Society, 1961, pp. 127–141; V. Lakshmikantham and S. Leela, *Differential and Integral Inequalities*, vols. I, II, Academic Press, New York, 1969.

conditions is given by the integral formula

(5) $$u(\mathbf{x}) = \int G(\mathbf{x}, \xi) f(\xi) \, d\xi.$$

Perhaps the oldest example of such a Green's function is $1/4\pi r$, the Green's function for the Poisson equation $-\nabla^2 u = f$. Historically, Poisson proceeded in the reverse order; he considered the "Newtonian potential" (5) of a mass (or charge) distribution of density $f(\mathbf{x})$ as defined by

$$u(\mathbf{x}) = \int [f(\mathbf{x})/|\xi - \mathbf{x}|] \, d\xi = \int G(\mathbf{x}, \xi) \, dm(\xi),$$

where $(\mathbf{x}, \xi) = 1/|\xi - \mathbf{x}|$ and $dm(\xi) = f(\xi) \, d\xi$; he then proved from this definition that $\nabla^2 u = f/4\pi$.

Another elementary example refers to

(6) $$-u'' = f(x), \quad u(-1) = u(1) = 0.$$

The Green's function for this is piecewise linear; it is

$$G(x, \xi) = -[|x - \xi| + x\xi - 1]/2.$$

Similarly, for $u^{iv} = f(x)$, with $u(0) = u'(0) = 0, u(1) = u'(1) = 0$ (a clamped beam), the Green's function is given by

$$12G(x, \xi) = |x - \xi|^3 - (x + \xi)^3 + 6x\xi(x + \xi)(1 + x\xi) - 4x^2\xi^2(3 + x\xi).$$

Note that the above Green's functions are *symmetric*, like $1/4\pi r = 1/4\pi|\mathbf{x} - \xi|$: we have in all cases $G(\mathbf{x}, \xi) = G(\xi, \mathbf{x})$. This symmetry is an important general property of the Green's function of *self-adjoint* DE's. Green's functions are also *positive* for boundary conditions of the form $\alpha u + \beta \partial u/\partial n = 0$ with $\alpha \geq 0, \beta \geq 0$ and $\alpha + \beta > 0$, for example, if α_0, α_1 and β_1 are positive and β_0 is negative in equation (3') of Lecture 1. As has been remarked by Kurt Friedrichs [Az, pp. 53–64], these two traits of symmetry and positivity are very general in mathematical physics.

From the Green's function, one can easily express the *Poisson kernel* (as its normal derivative on the boundary); specifically, in three dimensions:

(7) $$u(\mathbf{x}) = \int_\Omega \frac{\partial G}{\partial \nu}(\mathbf{x}, \xi) f(\xi) \, d\xi, \quad \text{if} \quad \nabla^2 u = 0.$$

For discs, spheres, and other special domains, the Poisson kernel $\partial G/\partial \nu$ can be expressed as an elementary function in closed form. In such cases, the easiest way to compute $u(\mathbf{x})$ for *one* $\mathbf{x}$ is from (7). However, if one wishes to tabulate a *field* of values, numerical methods become more efficient.

We shall discuss such classical integral transform and related integral equation methods further in Lecture 6, §4 and following sections. It should be stressed that they become impractical for elliptic DE's with variable coefficients in general domains.

5. Variational principles.

The solutions of many boundary value problems of mathematical physics minimize an appropriate functional. In particular, every configuration of static equilibrium in classical (Lagrangian) mechanics minimizes a suitable energy function.

Such variational principles (to be used in Lecture 8) include, for example,

$$(8) \qquad \delta \int [\tfrac{1}{2}p\nabla u \cdot \nabla u + q(\mathbf{x})u^2 - f(\mathbf{x})u(\mathbf{x})]|d\mathbf{x}| = 0.$$

For given boundary values, the integral is minimized by the solution of the source problem $\nabla \cdot (p\nabla u) = q(\mathbf{x})u - f(\mathbf{x})$. In the special case $p = 1$ and $q = f = 0$, this reduces to the Dirichlet principle.

A more sophisticated variational principle refers to a simply supported plate with Poisson ratio v and load density $\rho(x, y)$, resting on the plane $z = 0$. Here the equilibrium condition is $\delta J = 0$, for

$$(9) \qquad J[u] = \int \int_R \{\tfrac{1}{2}[(\nabla^2 u)^2 - 2(1 - v)(u_{xx}u_{yy} - u_{xy}^2)] - \rho u\} \, dx \, dy.$$

Surprisingly, the Euler–Lagrange DE for the minimum of (2) is $\nabla^4 u = \rho(x, y)/D$ regardless of v: the integral of $(u_{xx}u_{yy} - u_{xy}^2)$ is absorbed into a "boundary term" (boundary stress) and v does not affect (9) when $\rho = 0$.

A third example is furnished by the Plateau problem of Lecture 1, §8: to find a "minimal surface" (surface of least area) spanning a given contour γ.

Rayleigh quotient. Likewise, the eigenfunctions of the generalized Helmholtz equation

$$(10) \qquad \nabla \cdot (p(\mathbf{x})\nabla u) + k^2 \rho(\mathbf{x})u = 0 \quad \text{in} \quad R,$$

subject to the boundary condition $u \equiv 0$ on ∂R, are those for which the Rayleigh quotient

$$(11) \qquad J[u] = \int \int \int (p\nabla u \cdot \nabla u) \, dR \Big/ \int \int \int \rho u^2 \, dR$$

has a stationary value.

6. Existence theorems.

We have become so accustomed to the idea that "source problems" of the form $-\nabla \cdot (p\nabla u) = f(\mathbf{x})$, $p(\mathbf{x}) > 0$, have unique solutions for given boundary values, that it is natural to accept this result as "physically obvious" from the interpretation of Lecture 1, §2 and §3. Actually, the result is very hard to prove rigorously, even in the simplest case ($p = 1, f = 0$) of the Dirichlet problem:

$$(12) \qquad \nabla^2 u = 0 \quad \text{in} \quad R, \quad u = g(\mathbf{y}) \quad \text{on} \quad \partial R.$$

Indeed, Dirichlet's first "proof" of this fact was fallacious. It was based on the true Dirichlet principle, which he discovered, that to satisfy (12) was equivalent (for "smooth" functions) to *minimizing the Dirichlet integral* of (8):

$$(13) \qquad \int_R (\nabla u \cdot \nabla u)\, dR = \int_R \left\{ \sum \left(\frac{\partial u}{\partial x_i}\right)^2 \right\} dR,$$

subject to the specified boundary condition $u = g(\mathbf{y})$ on ∂R. This incidentally implies uniqueness, like the maximum principle of § 3. He also observed that, since the Dirichlet integral (9) is positive (or zero in the trivial case $u = $ const.), its values must have a greatest lower bound.

However, it is very hard to prove that there *exists* a (smooth) function which actually attains this minimum possible value. The "direct variational approach" to proving existence theorems is so far limited to the case of two space dimensions[7]; so are various function-theoretic methods [2, pp. 254–281]. For the Dirichlet problem, one can prove existence in n dimensions by the Poincaré–Perron "méthode de balayage," which is closely related to the relaxation methods which I shall discuss in Lecture 4. Using integral equation methods, one can prove from this the existence of solutions to the Laplace equation for other boundary conditions (see [K, Chap. XI]).

But there are regions in space and continuous boundary values for which the Dirichlet problem does *not* have a solution (e.g., the Lebesgue spine of [K, p. 285]). Moreover, the general existence theory is extremely technical, even for *linear* elliptic DE's (with variable coefficients). To have even a faint glimmering of what is known, one must begin by distinguishing "weak" from "strong" solutions.[8] For *nonlinear* elliptic problems (even the Plateau problem), only weak existence theorems are known.

One should also realize that many techniques for proving existence theorems are "constructive" only in a remote sense. Our main concern below will be with a very different problem: that of *effectively constructing numerical tables* of functions defined as solutions of specified elliptic boundary values and eigenvalue problems.

7. Smoothness. A characteristic property of elliptic equations, which is not shared by partial DE's of other types, is the fact that their solutions are as smooth as their coefficient-functions. For example, if an elliptic DE has analytic coefficient-functions, then all its solutions are analytic at all interior points [2, p. 136].

This smoothness makes it much easier to compute solutions of elliptic equations accurately by numerical methods, using either difference or variational approximations.

Moreover, solutions of analytic elliptic problems are also analytic on the boundary where the boundary and boundary data are analytic. Such solutions can therefore be continued across the boundary into an open region containing the

---

[7] See [3] for a good exposition and historical analysis; I shall discuss the numerical implementation of this approach in Lecture 8.

[8] See [2] for a classical exposition, or [6] for an authoritative recent survey.

boundary in its interior. This is because of various reflection principles, of which the following is the most famous and familiar.

Schwarz reflection principle. Let $w = f(z)$ be an analytic function of the complex variable z in the upper half-plane $y \geq 0$, and let w be real on the segment of the real axis $y = 0$. Then f can be continued analytically into the lower half-plane by setting $w = [f(z^*)]^*$ there, where z^* designates the complex conjugate of z.

Singularities at corners. Although solutions of elliptic DE's with analytic coefficients are analytic at interior points, and solutions of elliptic boundary value problems with analytic coefficients *and* boundary data are analytic where the boundary is analytic, such solutions generally have singularities at corners, edges, etc. Thus the solution of the Poisson equation $-\nabla^2 u = 1$ in the unit square S for the boundary condition $u \equiv 0$ on ∂S has a singularity at each corner; so does the deflection of a uniformly loaded plate: the solution of $-\nabla^4 u = 1$ in S with $u \equiv 0$ on ∂S [Az, p. 229].

Hence a priori truncation error estimates which involve differentiability assumptions are often inapplicable near corners and other points where boundaries are nonanalytic (cf. Lecture 6, § 3).

Using the Schwarz reflection principle, H. Lewy and his students have found the most general form of such singularities for a fairly wide family of conditions. One of the most powerful results is the following.

LEHMAN'S THEOREM.[9] *Let $u(x, y)$ be a solution of $\nabla^2 u + a(x, y)u_x + b(x, y)u_y = s(x, y)$ in an analytic corner subtending an angle $\alpha\pi$, and assuming analytic boundary values (except at the vertex). Then $u(x, y)$ is asymptotic to an analytic function of $z, z^*, z^{1/\alpha}, z^{*1/\alpha}$ if α is irrational, and of these variables and $z^q \log z, z^{*q} \log z^*$ if $\alpha = p/q$ is rational (in lowest terms).*

This result has been extended by Neil Wigley to the case of sufficiently smooth (i.e., differentiable) boundaries and boundary values.

For earlier work on singularities of elliptic equations, see [F, Chap. 24], and the references given there.

REFERENCES FOR LECTURE 2

[1] DOROTHY L. BERNSTEIN, *Existence Theorems in Partial Differential Equations*, Ann. of Math. Studies no. 23, Princeton University Press, Princeton, New Jersey, 1950.
[2] L. BERS, F. JOHN AND M. SCHECHTER, *Partial Differential Equations*, Wiley–Interscience, New York, 1964.
[3] R. COURANT, *Dirichlet's Principle*, Interscience, New York, 1950.
[4] PH. FRANK AND R. VON MISES, *Differentialgleichungen ... der Mechanik und Physik*, vols. I, II, Vieweg, 1930.
[5] P. R. GARABEDIAN, *Partial Differential Equations*, John Wiley, New York, 1964.
[6] C. B. MORREY, *Differentiability theorems for weak solutions of nonlinear elliptic differential equations*, Bull. Amer. Math. Soc., 4 (1969), pp. 684–705.
[7] N. I. MUSHKELISHVILI, *Some Fundamental Problems of the Theory of Elasticity*, 4th ed., Akademie-Verlag, Moscow, 1954.

---

[9] R. S. Lehman, J. Math. Mech., 8 (1959), pp. 727–760. For Wigley's extension, see Ibid., 13 (1964), pp. 549–576; 19 (1969), pp. 395–401.

[8] M. NICOLESCO, *Les Fonctions Polyharmoniques*, Hermann, Paris, 1936.
[9] M. H. PROTTER AND H. F. WEINBERGER, *Maximum Principles in Differential Equations*, Prentice-Hall, Englewood Cliffs, New Jersey, 1967, Chap. 3.
[10] A. N. TYCHONOV AND A. A. SAMARSKII, *Partial differential equations of mathematical physics*, vols. I, II, Holden-Day, San Francisco, 1964.

LECTURE 3

Difference Approximations

1. Five-point difference equation. As was stated in Lecture 2, §1, numerical methods have the great advantage of being applicable in principle to linear partial DE's with *variable* coefficients on *general domains* (but see §6). This is because one can approximate such DE's by difference equations (ΔE's); the present lecture will be devoted to a discussion of such approximating ΔE's. As was explained in the preface, the resulting "difference methods" (whose discussion will occupy Lectures 3–5) were used almost exclusively to solve elliptic problems on computers until recently.

We begin with the special case of the Laplace DE. It is classic that, knowing the values of a function $u \in C^4(R)$ at the mesh-points $(x_i, y_j) = (ih, jh)$ of a uniform square mesh, the Laplacian of u is approximated with $O(h^2)$ accuracy by the second central difference quotient

$$(1) \qquad \nabla_h^2 u(x_i, y_j) \doteq \frac{1}{h^2}[u_{i+1,j} + u_{i-1,j} + u_{i,j+1} + u_{i,j-1} - 4u_{i,j}].$$

Clearly, $\nabla_h^2 u = (\delta_x^2 + \delta_y^2)u/h^2$, where δ_x^2 and δ_y^2 signify second central difference operators.

To compute the *truncation error* $\nabla_h^2 u - \nabla^2 u$, we assume $u \in C^6(R)$ and expand in Taylor series, getting

$$(2) \qquad h^2 \nabla^2 u = (\delta_x^2 + \delta_y^2)u - (\delta_x^4 + \delta_y^4)u/12 + O(h^6).$$

Hence, dividing by h^2 and noting that $\delta_x^4 u + \delta_y^4 u = O(h^4)$, we see that the truncation error in (1) is $O(h^2)$ for any $u \in C^4(R)$.

A function which satisfies $\nabla_h^2 u = 0$ on a uniform mesh is called a "discrete harmonic function" (see §2). This is evidently equivalent to the condition that its value at each interior mesh-point is the *arithmetic mean* of its values at the four adjacent mesh-points.

More generally, consider the *source problem* in a bounded plane region R. As in Lecture 1, §3, this problem consists in solving the self-adjoint elliptic DE

$$(3) \qquad -\nabla \cdot [p(x,y)\nabla u] + q(x,y)u = s(x,y), \qquad p > 0, q \geq 0,$$

for suitable boundary conditions. This DE may be approximated at the interior mesh-points of any rectangular mesh by the following five-point central *difference equation* (ΔE):

$$(4) \qquad D_{i,j}u_{i,j} = L_{i,j}u_{i-1,j} + R_{i,j}u_{i+1,j} + T_{i,j}u_{i,j+1} + B_{i,j}u_{i,j-1} + q_{i,j}u_{i,j} + S_{i,j},$$

where

(4')
$$D_{i,j} = R_{i,j} + L_{i,j} + T_{i,j} + B_{i,j}, \quad \text{with}$$
$$R_{i,j} = -[p(x_{i+1}, y_j) + p(x_i, y_j)]/2(x_{i+1} - x_i);$$

and $L_{i,j}$, $T_{i,j}$ and $B_{i,j}$ are given by similar formulas [FW, p. 201], [V, p. 186]. As we shall see in § 5, the error in the approximation (4) to the DE (3) is $O(h^2)$ if a uniform mesh is used, but only $O(|\Delta x_i|)$ if a nonuniform mesh is used. That is, an order of accuracy is lost in passing from the case of *constant* mesh-length to the general case of *variable* mesh-length.

These approximations "reduce" the analytical source problem defined by the DE (3) and the Dirichlet boundary condition $u(x, y) = f(x, y)$ on ∂R, the boundary of R, to an approximately equivalent *algebraic* problem of solving a system of n simultaneous linear equations in n unknowns, where n is the number of interior mesh-points. In vector notation, this algebraic problem consists in solving a vector equation of the form

(5) $$A\mathbf{u} = \mathbf{b}.$$

Here $\mathbf{u}$ is the vector of unknown values of the $u_{i,j}$ at interior mesh-points, while A and $\mathbf{b}$ (the vector of boundary values and source terms) are known.

To solve large systems (5) of simultaneous linear equations efficiently and accurately is not easy; techniques for doing this when A is a $10^4 \times 10^4$ matrix, say, will be the main theme of my next two lectures. The success of such techniques depends basically on a number of special properties of A—and especially on the fact that A is a Stieltjes matrix whose off-diagonal entries form a 2-cyclic matrix, in the sense of the following definitions.

DEFINITION. A *Stieltjes matrix* is a symmetric matrix whose diagonal elements are positive, whose off-diagonal elements are negative or zero, and which is positive definite [V, p. 85]. An $n \times n$ matrix B is *2-cyclic* (or has "Property A") when its indices can be partitioned into two nonvoid subsets S and T such that $b_{kl} \neq 0$ implies $k \in S$ and $l \in T$ or vice versa.

We shall now verify that the square matrix $A = \|a_{kl}\|$ of coefficients[1] of the system (5) is *symmetric*, since

$$R_{i,j} = L_{i+1,j} = c_{i+1/2,j}, \quad T_{i,j} = B_{i,j+1} = c_{i,j+1/2}.$$

Next, decompose A as follows:

(6) $$A = D - E - F, \quad F = E^T,$$

where D is its diagonal component, $-E$ its subdiagonal component, and $-F$ its superdiagonal component. All three matrices D, E, F are nonnegative. Moreover, in particular, (i) A has positive diagonal entries and negative or zero off-diagonal entries ($-A$ is "essentially nonnegative"); also (ii) A is *diagonally dominant*, in the sense that each (positive) "diagonal" coefficient $D_{i,j}$ in (4) is equal to the sum

---
[1] Here each index (k or l) stands for a mesh-point (i, j).

of the magnitudes of all the other coefficients in its row, and greater than this sum for points where $q > 0$, and adjacent to points satisfying boundary conditions $u = g(\mathbf{x})$ or $\partial u/\partial n + b(\mathbf{x})u = g(\mathbf{x})$, $b(\mathbf{x}) > 0$; (iii) A is *positive definite*, i.e., $\mathbf{x}A\mathbf{x} > 0$ unless $\mathbf{x} = 0$; and (iv) it is a *Stieltjes matrix* with strictly positive inverse; finally, (v) A is *sparse* in that it has at most 4 nonzero off-diagonal entries in each row, and (vi) it is *2-cyclic* since one can take for S those $k = (i,j)$ with $i + j$ odd, and for T the $k = (i,j)$ with $i + j$ even. For a sufficiently fine mesh in a connected domain with smooth boundary, (vii) the matrix A is also *irreducible*.

2. Network analogies. Solutions of linear systems having a Stieltjes coefficient-matrix are of interest not only as approximate solutions of problems of continuum physics; they also represent exact solutions to interesting *network problems* arising in various branches of physics.

Specifically, let $C = \|c_{kl}\|$ be any Stieltjes matrix. We can construct a D.C. network whose kth node is connected with its lth node by a wire of conductance $-c_{kl}$ if $c_{kl} < 0$, and is not connected with node l when $c_{kl} = 0$ (i.e., otherwise). We let each jth node have an input lead with controlled current S_j and a resistive connection to "ground" with conductance $c_{jj} - \sum_k c_{jk} \geq 0$. Then Kirchhoff's laws are equivalent to the vector equation

(7) $$S_j = \sum_k c_{jk}(u_j - u_k) = \sum I_{jk} = -\sum I_{kj}.$$

The sparseness of C is reflected in a sparseness of links.

By inspection, we find that the 5-point difference approximation (4) to $-\nabla \cdot (p\nabla u) = f(x, y)$ leads to a rectangular D.C. network, whose nodes are the mesh-points and whose conducting elements are the mesh-segments. In this *network analogy*, $u_{i,j}$ is the voltage at the terminal (i,j), $R_{i,j} = L_{i+1,j} = c_{i+1/2,j}$ in (6) is the conductance of the wire connecting node (i,j) to node $(i + 1,j)$, $s_{i,j}$ is the current flowing into node (i,j), and so on. As a result of this analogy, one can build rectangular networks for solving the difference equations (4) by analogy. (Similarly, one can use an electrolytic tank or telegraphic "teledeltos" paper as analogue computer to solve the DE $\nabla^2 u = 0$.)

A mechanical analogy is provided by locating taut strings under constant tension T on the mesh-lines of a rectangular network, loaded at the mesh-points where these lines are joined, and looking for static equilibrium (minimum strain energy): the stationary state of minimum strain energy, with $\delta J = 0$ for

(8) $$J = \tfrac{1}{2}(\mathbf{u}, A\mathbf{u}) - \mathbf{b} \cdot \mathbf{u}.$$

This analogy suggested to Hardy Cross and to R. V. Southwell the idea of solving the resulting equations (i.e., of minimizing J) by iterative "relaxation" methods to be described in Lectures 4 and 5, in which J is repeatedly reduced by changing one $u_{k,l}$ at a time.

Though using variable mesh-length and nonrectangular meshes ("irregular stars" [9]) improve the accuracy of the network analogy in regions where the exact solution is rapidly varying, they also greatly complicate the writing of computer programs for solving the resulting systems of linear algebraic equations.

Discrete harmonic functions. The difference approximation (1) on a uniform mesh also defines a fascinating class of "discrete harmonic functions" $u(i,j)$, defined as solutions of the ΔE

$$(9) \qquad u(i,j) = \tfrac{1}{4}[u(i+1,j) + u(i,j+1) + u(i-1,j) + u(i,j-1)].$$

Discrete harmonic functions have been extensively studied by Duffin and others.

Note that (9) is analogous to Gauss' theorem of the arithmetic mean. Again, solutions of (9) minimize the "Dirichlet sum" $\sum (\Delta u)^2$ of the squares of difference ("jumps" in u between adjacent mesh-points) for given boundary values.

3. Solution by elimination. When the number N of mesh-points is moderate (when $N < 1000$, say), it is usually feasible to solve the system of ΔE's $A\mathbf{v} = \mathbf{k}$ by Gaussian elimination in single-precision arithmetic. However, this involves approximately $N^3/3$ multiplications [FW, Chap. 25], as well as storing up to $N^2 \sim 10^6$ numerical coefficients. The situation is very different from that in the one-dimensional case, in which the 3-point $O(h^2)$ approximation leads to only a *tridiagonal* matrix.

If the number of mesh-points on any horizontal line is bounded by M, then the matrix A is a *band matrix* with bandwidth at most $2M + 1$. Gaussian elimination then requires only about $M^2 N$ multiplications.[2]

Alternatively, one can regard the ΔE (1) (for example) as a *two-endpoint problem* for a two-level system of M second order ΔE's:

$$(10) \qquad u_{j+1}(i) = 4u_j(i) - u_{j-1}(i) - u_j(i+1) - u_j(i-1),$$

which can then be integrated using "multiple shooting" techniques. These have been studied by H. B. Keller[3] and others. However, the DE (10) is unstable, and this approach may well lead to a need for double precision [FW, loc. cit.].

Optimal elimination. Reduction to minimum bandwidth is only one of several techniques which have been developed for exploiting the sparseness of matrices arising from ΔE's and network problems. Reduction to minimum bandwidth does not always minimize the work of achieving exact solutions (in "rational arithmetic"): it is by no means always optimal. Indeed, the whole subject of optimizing elimination for sparse matrices is currently a very active research area; I can only give you a few major references.[4]

---

[2] For more details, see G. E. Forsythe and C. B. Moler, *Computer Solution of Linear Algebraic Systems*, Prentice-Hall, Englewood Cliffs, New Jersey, 1967, pp. 115–119.

[3] *Two-Endpoint Problems*, Blaisdell, Waltham, Massachusetts, 1968.

[4] See D. V. Steward, SIAM J. Numer. Anal., 2 (1965), pp. 345–365; R. A. Willoughby, editor, *Sparse Matrix Proceedings*, RA-1, IBM Res Publ., March, 1969; [4], [8], and Part D of my article to appear in Proc. SIAM–AMS Symp. IV (1971).

A very special elimination method, which is brilliantly successful for solving the Poisson equation $-\nabla^2 u = f$ in rectangular regions is the Tukey–Cooley fast Fourier transform on a uniform $2^m \times 2^n$ mesh.[5]

However, for most very large problems ($N > 10,000$, say) in general regions, and especially for those which involve multiple interfaces such as occur in nuclear reactors, stable and self-correcting *iterative* methods seem to be preferable. My next two lectures will be largely devoted to iterative and semi-iterative methods for solving large systems of simultaneous linear equations. These have the further advantage of being more readily adaptable to nonlinear problems.

4. Nonlinear problems. By the simple device of replacing derivatives by (approximately equal) difference quotients, *nonlinear* DE's can also be approximated by (nonlinear) systems of algebraic equations.

Methods for solving the resulting systems of nonlinear equations are typically iterative, beginning with *Newton's method*, which is the method most commonly proposed in textbooks. (The usual expositions of this method take for granted the triviality of solving *linear* systems, incidentally.)

For this reason, I shall postpone the study of (iterative) methods for solving systems of nonlinear algebraic equations to Lecture 4 (and to Lecture 8, § 3); their success for large systems usually depends on quite special considerations.

Nonlinear networks. For example, they may depend on variational properties, such as hold for a wide class of nonlinear networks[6] analogous to the linear networks discussed in § 2. From this principle, one can derive existence and uniqueness theorems for flows.

5. Local truncation errors. For the rest of this lecture, I shall ignore the practical difficulties of solving accurately large systems of algebraic equations, and describe what is known about the *accuracy* of difference approximations, assuming that the difference equations can be solved.

As I said in § 1, the 5-point central difference quotient approximation for $\nabla^2 u$ on a uniform mesh introduces an error of $O(h^2)$ at each mesh-point. Unfortunately, its generalization (4) to a nonuniform mesh (or even with a uniform mesh unless (3) has constant coefficients) introduces an error of $O(h)$. Moreover, this order of accuracy is "best possible": with only five mesh-points, one cannot match more than the five coefficients corresponding to u, u_x, u_y, u_{xx} and u_{yy} in the Taylor series expansion of u. It is sheer luck when other derivatives have no influence. Indeed, one cannot express u_{xy} even approximately in terms of the 5 values of u in (3). For this reason, difference approximations to elliptic problems in which

---

[5] R. W. Hockney, J. Assoc. Comput. Mach., 12 (1965), pp. 95–113; F. W. Dorr, SIAM Rev., 12 (1970), pp. 248–263; B. L. Buzbee et al., SIAM J. Numer. Anal., 7 (1970), pp. 623–656.

[6] G. Birkhoff and J. B. Diaz, Quart. Appl. Math., 13 (1956), pp. 432–443; see also G. Birkhoff and R. B. Kellogg, Proc. Symp. Generalized Networks, MRI Symposium Series 16, Brooklyn Polytechnic Press, New York, 1966, and the references of Lecture 8, footnote 8.

u_{xy} enter normally use a 9-point formula since, for example,

(11) $\quad h^2 u_{xy} = u_{i+1,j+1} + u_{i-1,j-1} - u_{i+1,j-1} - u_{i-1,j+1} + O(h^4).$

Instead of using truncated Taylor series to derive difference approximations to derivatives, one can use integral formulas. Careful discussions of this approach may be found in [FW], [KK], [V] and [W].

In either case, the most useful fact to be deduced from such a priori error estimates is the principle that the error (for a uniform mesh) is typically asymptotic to $Mh^n + O(h^{n+1})$ for some positive integer n.

Order of convergence. For square meshes with mesh-length h, the truncation error is typically of the form $Mh^n + O(h^{n+1})$ for some positive integer n, the "order of convergence." More generally, this is true of rectangular meshes with mesh-length $h\theta_k$ in the x_k-direction, and in many other cases. In such cases, the changes $\Delta u_{i,j}$ in computed values when the mesh-length is halved from h to $h/2$ are approximately proportional to h^n. Though M is unknown, one can use Richardson's method of "deferred approach to the limit" [FW, p. 307] to improve the accuracy of results obtained by mesh-halving (until roundoff takes over). See also [11].

6. Higher order accuracy. One can always approximate difference quotients of very smooth functions with higher order accuracy by using stencils with enough mesh-points; this follows from Taylor's formula. In the case of partial DE's with constant coefficients and a uniform mesh, the process yields some very elegant (and sometimes useful) formulas. I shall mention a few such formulas, giving references[7] and assuming high order differentiability.

Thus, formula (2) leads to a difference approximation

$$\nabla^2 u = \frac{1}{6h^2}[4\delta^2 u + \tilde{\delta}^2 u] + O(h^4),$$

where

$$\tilde{\delta}^2 u = [u_{i+1,j+1} + u_{i+1,j-1} + u_{i-1,j+1} + u_{i-1,j-1} - 4u_{ij}]$$

having $O(h^4)$ accuracy on a 9-point *square* of mesh-points (see [KK, p. 179] and J. Bramble and B. Hubbard [2]). This is not to be confused with the difference approximation

$$\nabla^2 u = \bar{\delta}_{xx} u + \bar{\delta}_{yy} u + O(h^4),$$

where

$$\bar{\delta}_{xx} u = [16(u_{i+1} + u_{i-1}) - (u_{i+2} + u_{i-2}) - 30u_0]/24h^2$$

on a 9-point *cross* of mesh-points [KK, p. 184], obtained by minimizing the Dirichlet integral on the piecewise bilinear function interpolated between values

---

[7] A useful compendium is contained in Collatz, Table VI [C, pp. 505–509]; see also W. G. Bickley et al., Proc. Roy. Soc. London, A262 (1961), pp. 219–236.

of u at these mesh-points.[8] This 9-point difference approximation with $O(h^4)$ accuracy applies also to DE's of the form $Au_{xx} + Cu_{yy} = 0$ and to

$$Au_{xx} + 2Bu_{xy} + Cu_{yy} + Du_x + Eu_y + Fu = 0$$

if $A = C$ or $B = 0$.[9]

One can obtain a difference approximation to $\nabla^2 u$ having $O(h^{10})$ accuracy by using a 13-point stencil [KK, p. 184], while 17-point stencils for $\nabla^2 u$ and 25-point stencils for $\nabla^4 u$ have also been worked out.[10]

Finally, accurate difference approximations to ∇^2 on triangular and hexagonal nets have been worked out by various authors.[11] Unfortunately, although such higher order methods are intriguing, the use of the associated larger stencils almost invariably leads to serious complications near the boundary.

7. Global error bounds. The errors referred to in § 1 and § 6 were discrepancies between difference quotients ("divided differences") and derivatives. The question arises: how are such errors related to those in the *values* of the functions? If we write the difference approximation in the form $A\mathbf{v} = \mathbf{k}$, then we have $A\mathbf{u} = \mathbf{k} + \mathbf{r}$, where $\mathbf{r}$ is the vector whose components r_i are these discrepancies. The r_i are also called *residuals* for the system $A\mathbf{u} - \mathbf{k}$.

For the Laplace equation, and in some other cases, one can achieve higher order (local) accuracy a posteriori by estimating the r_i from numerical data (e.g., by estimating $\nabla^4 \mathbf{u}$ from the computer printout). If $\bar{\mathbf{r}}$ is the estimated dominant error term, then by subtracting the solution of $A\mathbf{v} = \bar{\mathbf{r}}$ as a "differential correction" from the solution of the difference approximation, one should reduce the error. This is Fox's "method of differential corrections" [5].[12]

Discrete Green's function. Alternatively, one can combine remainder formulas with a priori knowledge of the derivatives of the exact solution, obtained by analytic considerations (cf. Lecture 2), to bound the residuals r_i. Since the actual error vector $\mathbf{e} = \mathbf{v} - \mathbf{u}$ satisfies $\mathbf{e} = G\mathbf{r}$ where $G = A^{-1}$, this leads to an *a priori error bound* in terms of the norm of G. Here G may be called the *Green's matrix* because it acts like a *discrete Green's function* [FW, pp. 315–318] for the source problem being solved. It is a positive matrix for (4). Finally, again using analytical considerations discussed in Lecture 2, one can often bound the norm of G.

---

[8] R. Courant, Bull. Amer. Math. Soc., 49 (1943), pp. 1–27; B. Epstein, Math. Comp., 16 (1962), pp. 110–112.

[9] J. Bramble and B. Hubbard, Contributions to Differential Equations, 2 (1963), pp. 319–340; Young and Dauwalder, Rep. TNN-46, Univ. of Texas Comp. Lab.

[10] B. Meister and W. Prager, Z. Angew Math. Phys., 16 (1965), pp. 403–410; see also G. Fairweather et al., Numer. Math., 10 (1967), pp. 56–66; A. Hadjimos, Ibid., 13 (1969), pp. 396–403; and F. D. Burgoyne, Math. Comp., 22 (1968), pp. 589–594.

[11] [KK, pp. 187–188]; R. B. Kellogg, Math. Comp., 18 (1964), pp. 203–210; [C]; [9]; D. N. de G. Allen, *Relaxation Methods in Engineering and Science*, McGraw-Hill, New York, 1954. I. Babuška, M. Prager and M. Vitasek, *Numerical Processes in DE's*, SNTL-Interscience, 1966, § 5.4.2.

[12] See also [Az, p. 203], and E. A. Volkov, Vychisl. Mat., 1 (1957), pp. 34–61 and 62–80.

Using such considerations, global convergence as $h \downarrow 0$ was first proved for the Laplace ΔE on a square mesh by R. G. D. Richardson in 1917 and by Phillips and Wiener in 1922; the aim of these authors was to establish *existence theorems* for solutions of the Dirichlet problem for $\nabla^2 u = 0$ from algebraic existence theorems for $\nabla_h^2 u = 0$. In 1927, Courant, Friedrichs and Lewy showed that all difference quotients of given order converged to the appropriate derivatives, as $h \downarrow 0$.

The maximum principle of Lecture 2, § 3, was applied to the Poisson equation by Gerschgorin [6] in 1930 to prove $O(h)$ global accuracy. Using linear interpolation on the boundary, Collatz[13] sharpened this result in 1933, under appropriate differentiability assumptions, to prove $O(h^2)$ accuracy. Further work was also done by Walsh and Young and by Wasow in 1954–5, and by P. Laasonen, who discussed carefully the loss of accuracy introduced by corners, where local singularities occur.[14] This literature is reviewed in [FW, p. 302], and in [C, pp. 326–327]. When mesh-points on the boundary are extremely close together, errors can be greatly magnified. A way to resolve this difficulty has been described by Babuška, Prager and Vitasek (op. cit., p. 274).

The whole subject was carefully reconsidered by Bramble and Hubbard, who used the Green's function approach systematically. They published their results in a series of papers written in 1964–5, especially in [1]–[2] and the references given there.[15] A significant question is whether or not A must be "monotone," i.e., whether the inverse G of A needs to be nonnegative. On this point, see [3] and recent work by Harvey Price.[16] The preceding authors have shown that, by using higher order differences, one can obtain higher order accuracy (for $\nabla^2 u = f$ and $\nabla^4 u = f$ on a square mesh).

The accuracy of the 5-point difference approximation with variable coefficients has been studied by Bramble, Hubbard and Thomée,[17] under weakened assumptions of smoothness on the boundary. For $u \in C^4(R) \cap C^2(\bar{R})$, for example, one obtains $O(h^2)$ accuracy. Finally, the $O(h^2)$ convergence of *all* difference quotients to the appropriate derivatives has been proved for the Laplace DE on a square mesh by V. Thomée and Achi Brandt.[18] Making increased smoothness assumptions, Thomée also showed that difference quotients converge at the same rate as the solution in the interior (giving discrete Harnack-type inequalities).

Many other more general results have been proved. Thus V. Thomée has proved convergence to order $O(h^{1/2})$ for simple difference approximations to the Dirichlet problem for any linear, constant-coefficient equation of elliptic type, and McAllister

---

[13] L. Collatz, Z. Angew Math. Mech., 13 (1933), pp. 56–57.

[14] See [7]; J. Assoc. Comput. Mach., 5 (1958), pp. 32–38; also E. Batschelet, Z. Angew Math. Phys., 3 (1952), pp. 165–193; N. M. Wigley, SIAM J. Numer. Anal., 3 (1966), pp. 372–383.

[15] Including Contributions to Differential Equations, 2 (1963), pp. 229–252; 3 (1963), pp. 319–340; SIAM J. Numer. Anal., 2 (1965), pp. 1–14; J. Assoc. Comput. Mach., 12 (1965), pp. 114–123; Numer. Math., 4 (1962), pp. 313–332; Ibid., 9 (1966), pp. 236–249.

[16] H. Price, Math. Comp., 22 (1968), pp. 489–516.

[17] BIT, 8 (1968), pp. 154–173. See also N. S. Bahalov, Vestnik Moskov Univ., 5 (1959), pp. 171–195, and J. R. Kuttler, SIAM J. Numer. Anal., 7 (1970), pp. 206–232.

[18] Math. Comp., 20 (1966), pp. 473–499. See also P. G. Ciarlet, Aequat. Math., 4 (1970), pp. 206–232.

has obtained global error bounds for difference approximations to certain mildly nonlinear elliptic problems.[19] Finally, Bramble [BV, pp. 201–209] has shown that by appropriately smoothing f, one can get improved convergence of difference approximations to $L[u] = f$, for uniformly elliptic L.

REFERENCES FOR LECTURE 3

[1] J. H. BRAMBLE AND B. E. HUBBARD, *Approximation of derivatives by difference methods in elliptic boundary value problems*, Contributions to Differential Equations, 3 (1964), pp. 399–410.

[2] ———, *New monotone type approximations for elliptic problems*, Math. Comp., 18 (1964), pp. 349–367.

[3] ———, *On a finite difference analogue of an elliptic boundary problem which is neither diagonally dominant nor of non-negative type*, J. Math. and Phys., 43 (1964), pp. 117–132.

[4] G. E. FORSYTHE AND C. B. MOLER, *Computer Solutions of Linear Algebraic Systems*, Prentice-Hall, Englewood Cliffs, New Jersey, 1967.

[5] L. FOX, *Some improvements in the use of relaxation methods for the solution of ordinary and partial differential equations*, Proc. Roy. Soc. London Ser. A, 190 (1947), pp. 31–59. (See also Philos. Trans. Roy. Soc. London. Ser. A, 242 (1950), pp. 345–378; Quart. J. Mech. Appl. Math., 1 (1948), pp. 253–280.)

[6] G. GERSCHGORIN, *Fehlerabschätzung für das Differenzenverfahren ...*, Z. Angew. Math. Mech., 10 (1930), pp. 373–382.

[7] P. LAASONEN, *On the degree of convergence of discrete approximations for the solutions of the Dirichlet problem*, Ann. Acad. Sci. Fenn. Ser. A, 246 (1957), 19 pp.

[8] A. RALSTON AND H. S. WILF, editors, *Numerical Methods for Digital Computers*, vol. II, John Wiley, New York, 1967.

[9] R. V. SOUTHWELL, *Relaxation Methods in Theoretical Physics*, Clarendon Press, Oxford, 1946.

[10] J. NITSCHE AND J. C. C. NITSCHE, *Error estimates for the numerical solution of elliptic differential equations*, Arch. Rational Mech. Anal., 5 (1960), pp. 293–306; pp. 307–314.

[11] V. PEREYRA, *Accelerating the convergence of discretization algorithms*, SIAM J. Numer. Anal., 4 (1967), pp. 508–533. (See also Numer. Math., 8 (1966), pp. 376–391, and 11 (1968), pp. 111–125.)

---

[19] V. Thomée, Contributions to Differential Equations, 3 (1964), pp. 301–324; G. T. McAllister, J. Math. Anal. Appl., 27 (1969), pp. 338–366.

LECTURE 4

Relaxation Methods

1. Point-Jacobi method. This lecture and the next will be devoted to *iterative* and *semi-iterative* methods for solving systems of linear equations (vector equations) of the form

(1) $$A\mathbf{u} = \mathbf{b}.$$

For very large systems involving 10^4 unknowns, these are usually more efficient than the elimination methods described in Lecture 3, §3 (see §7).

A great variety of such methods have been proposed; those involving "relaxation methods" are especially applicable when A is a *Stieltjes matrix* of the form

(2) $$A = D - E - F, \quad F = E^T.$$

As was shown in Lecture 3, §2, such matrices arise naturally from D.C. *network problems*, including those which correspond to the 5-point difference approximation to a source problem (with or without leakage). As we saw in that section, they also arise from the usual difference approximation to *second order* self-adjoint elliptic DE's of the form $-\nabla \cdot (p\nabla u) + qu = f$. When applied to such problems, many iterative methods are suggested by concepts of *relaxation* or *overrelaxation*, which may be motivated as follows.

The solution of (1) is that (column) vector $\mathbf{u}$ which minimizes the (positive definite) quadratic functional

(3) $$J(\mathbf{u}) = \tfrac{1}{2}\mathbf{u}^T A\mathbf{u} - \mathbf{b} \cdot \mathbf{u}.$$

In the loaded membrane physical interpretation, this functional is just the total *potential energy* of the system.

Example 1. For the 5-point discretization of $-\nabla^2 u = f(\mathbf{x})$, the functional to be minimized is the sum

$$\tfrac{1}{2}\sum [(u_{i,j} - u_{i,j-1})^2 + (u_{i,j} - u_{i-1,j})^2] + \sum f_{i,j} u_{i,j}.$$

As was explained in §2 of the previous lecture, one may simplify interpretation of relaxation methods by thinking of $J(\mathbf{u})$ as the "strain" energy of a configuration $\mathbf{u}$, whose coordinates u_j are "relaxed" cyclically so as to reduce J at each step. Finding this minimum by successively "relaxing" components u_j at $\mathbf{u}$, so as to reduce $J(\mathbf{u})$, is a simple way of looking at relaxation methods. This was also the idea of Poincaré's "méthode de balayage" for solving the Dirichlet problem, which likewise reduces the Dirichlet integral at each sweep.[1]

If one "scales" the quantities b_i, replacing them by $d_i^{-1} b_i = k_i$, the equation (1) is premultiplied by the diagonal matrix D^{-r}, which transforms it to the equivalent vector equation
$$u = D^{-1}(E + F)u + D^{-1}\mathbf{b}.$$

This suggests the iterative process

(4) $$\mathbf{u}^{(n+1)} = D^{-1}(E + F)\mathbf{u}^{(n)} + D^{-1}\mathbf{b},$$

which *is* the *point-Jacobi* method, also called the "method of simultaneous displacements."

If A is a Stieltjes matrix, the point-Jacobi method always converges [V, Theorems 3.3 and 3.6]. In particular, it converges for the matrix problems associated with any connected (irreducible) network, except when the current is specified at all boundary nodes, and there is no leakage.

2. Rate of convergence. Not only the fact of convergence but the *rate* of convergence is of crucial importance for an iterative method. For (4), this depends on the spectrum of $D^{-1}(E + F)$, which may also be written as

$$B = D^{-1}(E + F) = D^{-1/2}[D^{-1/2}(E + F)D^{-1/2}]D^{-1/2}.$$

In this notation, (4) simplifies to

(5) $$\mathbf{u}^{(n+1)} = B\mathbf{u}^{(n)} + \mathbf{k}, \quad B = D^{-1}(E + F), \quad \mathbf{k} = D^{-1}\mathbf{b}.$$

Since B is similar to a symmetric matrix $D^{-1/2}(E + F)D^{-1/2}$ (is "symmetrizable"), all eigenvalues of B are *real*.

In general, the matrix A underlying the point-Jacobi method for any well-designed difference approximation to a *self-adjoint* elliptic boundary value problem should be *symmetric*. Hence it and B should be similar to a *real diagonal* matrix.

Spectral radius. We now consider in some detail the questions of the *convergence* and the asymptotic *rate of convergence* of the point-Jacobi iterative method (4). The relevant concept is the *spectral radius* of B, $\rho(B)$. This is defined as the maximum of the magnitudes (absolute values) of the eigenvalues λ_i of B: $\rho(B) = \max |\lambda_i(B)|$.

By considering the Jordan canonical form $J = PBP^{-1}$ of B (P nonsingular), which is real and diagonal in the present case, it is easy to prove that (5) gives for any $\mathbf{u}^{(0)}$ a sequence of $\mathbf{u}^{(n)}$ which converge as $n \to \infty$ to the (unique) solution $\mathbf{u}$. In fact, the *error* $\mathbf{e}^{(n)} = \mathbf{u}^{(n)} - \mathbf{u}$ satisfies $\mathbf{e}^{(n)} = B^n \mathbf{e}^{(0)}$. When $\rho(B) < 1$, the norm of the error thus tends to zero, asymptotically like $[\rho(B)]^n$. Hence, the *asymptotic rate of* convergence as $n \to \infty$ is asymptotically proportional to $-\log \rho(B)$ if $\rho(B) < 1$;[2] if $\rho(B) \geq 1$, the method fails to converge.

---

[1] H. Poincaré, Amer. J. Math., 12 (1890), pp. 216–237; [K, p. 283].

[2] In the sense that, asymptotically, the error decreases by a factor e every $1/(-\log \rho(B))$ iterations (cf. [FW, p. 218]).

Remark. In some cases, one can interpret (4) as the Cauchy polygon method for integrating $\mathbf{u}_t = -A\mathbf{u} + \mathbf{k}$ with a small time step (see [V, § 8.4]). Thus, this is true for the usual 5-point approximation to $-\nabla^2 u = f$ on a uniform mesh; in this case, the point-Jacobi method (4) gives the Schmidt process for integrating the heat equation with source, $u_t = \nabla^2 u + f$.

More generally, all eigenvalues of A are *positive* for suitable mathematical models of most source problems, "passive" D.C. electrical networks, and other conservative or dissipative physical systems in the linear (small amplitude) range, including those of elasticity. Hence, $\mathbf{u}^{(r+1)} = \mathbf{u}^{(r)} + \Delta t(A\mathbf{u}^{(r)} - \mathbf{k})$ is convergent for sufficiently small Δt. If one takes the eigenvectors of A for coordinate axes, one can interpret (4) as integrating the system $dv_i/dt = -\lambda_i v_i + g_i$, where the λ_i are the (positive) eigenvalues of A.

The optimum Δt depends on the ratio $\lambda_{\max}/\lambda_{\min}$. In general, this is not easy to estimate, but see § 4.

3. Gauss–Seidel method. The point-Jacobi method yields every component of $\mathbf{u}^{(n+1)}$, the $(n+1)$st approximation to the solution vector of (1), as a (linear) function of components of $\mathbf{u}^{(n)}$, which are nearby in the case of difference schemes.

Thus, for the 5-point approximation to the Laplace DE, the point-Jacobi scheme at interior mesh-points is

(6) $$u_{i,j}^{(n+1)} = \tfrac{1}{4}[u_{i+1,j}^{(n)} + u_{i-1,j}^{(n)} + u_{i,j+1}^{(n)} + u_{i,j-1}^{(n)}].$$

Alternatively, sweeping through the components cyclically, one can *use improved values as soon as available*. Thus, for the natural ordering of mesh-points, one can use

(7) $$u_{i,j}^{(n+1)} = \tfrac{1}{4}[u_{i+1,j}^{(n)} + u_{i-1,j}^{(n+1)} + u_{i,j+1}^{(n)} + u_{i,j-1}^{(n+1)}].$$

The resulting method is called the *Gauss–Seidel method* (also the method of "successive displacements"). In general, the (point) Gauss–Seidel method is defined for (4) by

(7′) $$\mathbf{u}^{(n+1)} = (D - E)^{-1} F \mathbf{u}^{(n)} + (D - E)^{-1} \mathbf{b}.$$

It requires only half as much storage as point-Jacobi.

Stein–Rosenberg theorem. A very general theorem, due to Stein and Rosenberg, asserts that the preceding Gauss–Seidel method converges at least as fast as the point-Jacobi method. The proof depends only on the fact that the iteration matrix B is *nonnegative* with *zero diagonal entries*; thus B need not be symmetrizable for it to apply.

In the *2-cyclic* case of § 5 (e.g., for (7)), Gauss–Seidel converges exactly *twice* as fast as Jacobi for any given B. This is a fundamental result of David Young [V, p. 107].

The stopping problem. With iterative methods, a basic question is when to stop iterating. Criteria may be given in terms of either $\|\mathbf{u}^{(n+1)} - \mathbf{u}^{(n)}\|$ (in any norm) or, better, of the residual $\|A\mathbf{u} - \mathbf{b}\|$ and the rate of convergence of the process. We shall not enter into this question, beyond noting that, for ill-conditioned matrices, roundoff can pose surprising problems with Gauss–Seidel iteration.

Thus Wilkinson (J. Assoc. Comput. Mach., 8 (1961)) takes

$$A = \begin{bmatrix} .96326 & -.81321 \\ -.81321 & .68654 \end{bmatrix}, \quad \mathbf{b} = \begin{bmatrix} .88824 \\ -.74968 \end{bmatrix},$$

with the initial trial $\mathbf{x}^{(1)} = \begin{bmatrix} 0 \\ -.7 \end{bmatrix}$. Then, to five decimal digits,

$$\mathbf{x}^{(2)} = \mathbf{x}^{(3)} = \cdots = \begin{bmatrix} .33116 \\ -.70000 \end{bmatrix},$$

yet $A^{-1}\mathbf{b} = \begin{bmatrix} .39473\ldots \\ -.62470\ldots \end{bmatrix}$. Professor Moler kindly called this example to my attention.

4. Rates of convergence. Although the spectral radius $\rho(B)$ plays a central role in the theory of the rate of convergence of iterative methods, it is very hard to compute accurately. In practice, $\rho(B)$ must usually be estimated from numerical experiments (see § 7). However, there are a few exceptional "model problems" in which not only $\rho(B)$ but the entire spectrum is known.

Example 2. The eigenfunctions of the Laplace and Poisson equations in the rectangle $[0, a] \times [0, b]$ are $\sin(j\pi x/a)\sin(k\pi y/b)$. If this rectangle is subdivided by a uniform mesh into $M \times N$ subrectangles, the values of any of the above eigenfunctions at mesh-points define an eigenvector for the 5-point difference approximation to $-\nabla^2$. The case of a square (and the approximation $-\nabla_h^2$) is typical; the eigenvalues (for $j = 1, \cdots, M - 1$ and $k = 1, \cdots, N - 1$) are

$$4[\sin^2(j\pi/2M) + \sin^2(k\pi/2N)];$$

they range from $\lambda_{\min} = 4[\sin^2(\pi/2M) + \sin^2(\pi/2N)]$ to $2 - \lambda_{\min}$. Those of B range from $\lambda_{\min} - 1$ to $1 - \lambda_{\min}$; hence the spectral radius of the corresponding point-Jacobi iteration matrix is

$$\rho(B) = \tfrac{1}{2}[\cos(\pi/M) + \cos(\pi/N)] = 1 - \sin^2(\pi/M) - \sin^2(\pi/N)$$

and the eigenvalues of A correspondingly are $\pi^2[(j/a)^2 + (k/b)^2]$, $j = 1, \cdots, M$, $k = 1, \cdots, N$. Hence the eigenvalues of A range from $\sin^2(\pi/M) + \sin^2(\pi/N) = \lambda_{\min}$ to $\sin^2[(M - 1)\pi/M] + \sin^2[(N - 1)\pi/N] = 2 - \lambda_{\min}$.

Similar formulas can be written whenever $A = A' \otimes A''$ is a tensor product of tridiagonal matrices: $\lambda_{kl}(A) = \lambda_k(A')\lambda_l(A'')$, and the eigenvalues of tridiagonal matrices can be estimated as in Example 2 of Lecture 2. Moreover, the eigenvalues of A depend monotonically on its coefficients and the domain, so that comparison theorems can be invoked.[3] Finally, in diffusion problems with absorption, when the diffusion length is only a few mesh-lengths (e.g., when $h^2\sigma/p > 0.1$, say, for the DE $\nabla(p\nabla u) - \sigma u = s(x, y)$), and more generally when D strongly dominates $E + F$ in (2), the spectral radius can be estimated from this fact alone.

---
[3] See P. R. Garabedian, Math. Tables Aid. Comput., 10 (1956), pp. 183–185.

Example 3. For a uniform mesh on a square and Dirichlet (clamped plate) boundary conditions, the eigenfunctions of ∇^4 and the fourth order central difference approximation $\nabla_h^4 = (\nabla_h^2)^2$ to it can again be found by inspection. Using the results of Example 2, we can verify that the eigenvalues of ∇_h^2 satisfy

$$\mu_{\min} = \lambda_{\min}^2 \leq \mu_k \leq 2 - \mu_{\min},$$

where $\lambda_{\min} = 2 \sin^2 h$, whence $\mu_{\min} = 4 \sin^2 h$. Although the matrix A associated with the operator ∇_h^4 is not a Stieltjes matrix, one can apply to it the block overrelaxation methods to be discussed in § 7.

5. Point SOR. More generally, we can define *point* SOR for any "relaxation factor" ω as follows (SOR is an acronym for "successive overrelaxation"):

(8) $$(D - \omega E)\mathbf{u}^{(n+1)} = \{(1 - \omega)D + \omega F\}\mathbf{u}^{(n)} + \omega \mathbf{b}.$$

Setting $L = D^{-1}E$ and $U = D^{-1}F$, this becomes

(8') $$\mathbf{u}^{(n+1)} = (1 - \omega L)^{-1}[(1 - \omega)I + \omega U]\mathbf{u}^{(n)} + \omega(1 - \omega L)^{-1}D^{-1}\mathbf{b}.$$

When this method is applied over a complete cycle of mesh-points, the errors are transformed linearly in conformity to the formula

(9) $$\mathbf{e}^{(n+1)} = (1 - \omega L)^{-1}\{(1 - \omega)I + \omega U\}\mathbf{e}^{(n)},$$

which we rewrite as

(9') $$\mathbf{e}^{(n+1)} = L_\omega[\mathbf{e}^{(n)}], \quad L_\omega = (1 - \omega L)^{-1}\{(1 - \omega)I + \omega U\}.$$

For Stieltjes matrices, the Ostrowski–Reich theorem [V, p. 77] asserts that $\rho(L_\omega) < 1$ (in other words, point SOR converges) if and only if A is positive definite and $0 < \omega < 2$.

Kahan's thesis. In his unpublished thesis, W. Kahan (1958) extended to general Stieltjes matrices, in less sharp form, many of the results on point SOR which had been obtained by Young for the 2-cyclic case. Specifically, he showed that Young's best optimal overrelaxation factor ω_b was still good. We summarize his results as follows (for details, see [V, Theorems 4.9 and 4.12]).[4]

Let $A\mathbf{x} = \mathbf{b}$, where A is a Stieltjes matrix. Then we can rescale the known b_i so as to get an equivalent system

(10) $$D^{-1}A\mathbf{x} = \mathbf{c}, \quad \mathbf{c} = D^{-1}\mathbf{b}, \quad D = \text{diag } A.$$

Though $D^{-1}A$ is of course similar to the Stieltjes matrix

$$D^{1/2}AD^{1/2} = D^{-1/2}(DA)D^{1/2},$$

it is not itself generally a Stieltjes matrix. Both $D^{-1}A$ and $D^{1/2}AD^{1/2}$ have 1's on the main diagonal. Now rewrite (1) in the form

(11) $$\mathbf{x} = B\mathbf{x} + \mathbf{c}, \quad B = I - D^{-1}A,$$

---

[4] We also thank David Young for the exposition abstracted here (personal communication).

most suitable for iteration. Let $\mu = \rho(B)$ be the spectral radius of B:

$$\mu = \rho(D^{1/2}AD^{1/2} - 1) < 1$$

by §4. Apply *successive point-overrelaxation* (point SOR) to (11), with the particular (over)relaxation factor

(12) $\qquad \omega_b = 2/(1 + \sqrt{1 - \mu^2}) = 1 + [\mu/(1 + \sqrt{1 - \mu^2})]^2.$

Kahan has proved that for this ω_b,

$$\omega_b - 1 \leq \rho(L_{\omega_b}) \leq \sqrt{\omega_b - 1};$$

hence this ω_b is a *good* relaxation factor, since $\rho(L_\omega) \geq \omega_b - 1$ for *any* relaxation factor. For $\rho(B) = 1 - \varepsilon$, where ε is small, the asymptotic convergence rate $\gamma = -\log \rho(L_{\omega_b})$ therefore satisfies

(13) $\qquad \sqrt{2\varepsilon} = -\tfrac{1}{2}\log(\omega_b - 1) \leq \gamma \leq \log(\omega_b - 1) = 2\sqrt{2\varepsilon}.$

Rates of convergence. By combining the preceding considerations with those of §4, one can show that the rate of convergence of SOR is $O(h)$ for second order and $O(h^2)$ for fourth order elliptic problems (see again [V], [9]).

Two-cyclic case. The original and simplest class of applications of point SOR was to the case of 5-point difference approximations to *self-adjoint* elliptic problems on a rectangular mesh. In this case, the matrix B for the point-Jacobi method is (weakly) 2-*cyclic*, in the sense that for an appropriate ordering of the entries (indices), it has the form sketched below:

$$B = \begin{bmatrix} O & C \\ \tilde{C} & O \end{bmatrix}.$$

This is most easily visualized by interpreting the nodes as forming a *checkerboard* of red and black squares, such that no link (nonzero entry of B) joins two squares of the same color. The matrix displayed above is obtained by listing first all red squares and then all black squares. Although this ordering gives Gauss-Seidel and SOR their optimal (minimum) spectral radius, it is complicated as regards transfer of data from tape to core; this is handled better by a straightforward row-by-row (or column-by-column) sweep of mesh-points.

However, the 2-cyclic form of B displayed can be made to yield a significant economy: it suffices to store values at red mesh-points during *even* cycles (half-iterations) and values at black mesh-points during *odd* cycles. In symbols, write $\mathbf{u} = \mathbf{v} + \mathbf{w}$, where $\mathbf{v}$ and $\mathbf{w}$ are the vectors whose components are the values of $\mathbf{u}$ at red and black mesh-points, respectively (see [V, p. 150]). Then $\mathbf{v}^{(n+2)} = \tilde{C}C\mathbf{v}^{(n)}$, and data transfer becomes efficient if one sweeps through all red mesh-points row-by-row, and then all black mesh-points row-by-row.

Optimum overrelaxation parameters. In the 2-cyclic case which he originally considered, Young gave an exact formula for the optimum overrelaxation factor ω_b:

(14) $\qquad \omega_b = 2/[1 + \sqrt{1 - \rho^2(B)}]$ $\hfill$ [V, p. 110]

and the asymptotic rate of convergence [V, p. 106], in terms of the optimum SOR parameter ω_b. The eigenvalues μ of point SOR are related to those λ of the point-Jacobi method by

(15) $$(\lambda + \omega - 1)^2 = \lambda \omega_b^2 \mu^2 \qquad [V, (4.18)].$$

They lie on a circle in the complex plane which is mapped 2–1 and conformally onto the slit of real eigenvalues of B.

To illustrate the effectiveness of the SOR method, consider the case when A is a 2-cyclic Stieltjes matrix and $\rho^2 = .9999$. For this problem the Gauss-Seidel method would require an average of 25,000 iterations (neglecting roundoff) to get an extra decimal place of accuracy, whereas the SOR method, using optimum ω, would require only 115 iterations. However, to achieve this rapid convergence the overrelaxation parameter ω, or equivalently ρ^2, must be estimated accurately. For the above example, if an estimate of .999 were used for ρ^2 in computing ω, the SOR method would require 704, instead of 115, iterations to reduce the error by a factor of ten. When ρ^2 is close to unity, small changes in the estimate for ρ^2 can drastically affect the rate of convergence, especially if ω is underestimated.

In practice, two different numerical schemes have been widely used to obtain estimates for ρ^2 (or equivalently ω_b). One approach is to attack the eigenvalue problem directly and calculate ρ^2 prior to starting the main SOR iterations (see, for example, [4]). The second approach is to start with the SOR iterations with some $\omega < \omega_b$ and then obtain new estimates for ω based on numerical results.[5] The second approach is of the "semi-iterative" type to be discussed in Lecture 5.

p-cyclic matrices. Varga has generalized many properties of 2-cyclic matrices (matrices having "Property A" in Young's terminology) to *p*-cyclic matrices, such as arise in the "outer iterations" of the multigroup diffusion equations [V, Chap. 4]. In particular [V, Theorem 4.5], the spectral radius of SOR is again $1 - O(h)$ with the optimum overrelaxation parameter ω_b. However, it is the case $p = 2$ which arises most frequently in applications.

6. Richardson's method. There are several variants of SOR which have approximately the same rate of convergence. One of these is the second order Richardson's method,[6] a two-step method of "simultaneous displacements" which expresses $\mathbf{u}^{(n+1)}$ in terms of $\mathbf{u}^{(n)}$ and $\mathbf{u}^{(n-1)}$. It has the disadvantage of requiring twice as much storage as SOR.

After setting $\mathbf{v}^{(0)} = \mathbf{B}\mathbf{u}^{(0)} + \mathbf{k}$, one can replace (1) by the following larger system [V, pp. 142–143]:

(16a) $$\mathbf{u}^{(n+1)} = \omega[\mathbf{B}\mathbf{v}^{(n)} + \mathbf{k} - \mathbf{u}^{(n)}] + \mathbf{u}^{(n)},$$

(16b) $$\mathbf{v}^{(n+1)} = \omega[\mathbf{B}\mathbf{u}^{(n+1)} + \mathbf{k} - \mathbf{v}^{(n)}] + \mathbf{v}^{(n)},$$

---

[5] See [V, Chap. 9]; Hageman and Kellogg [4]; and J. K. Reid, Comput. J., 9 (1966), pp. 200–204.

[6] L. F. Richardson, Philos. Trans. Roy. Soc. London Ser. A, 210 (1910), pp. 307–357; see [V, p. 159]. Richardson did not use the 2-cyclic concept.

which is *2-cyclic* even if B is not. For the choice $\omega = \omega_b = 2/[1 + (1 - \rho^2(B))^{1/2}]$, one achieves a rate of convergence which is about half the optimal rate, as in Kahan's analysis.

SSOR overrelaxation. Another variant of SOR is Sheldon's "symmetric" SOR (or SSOR), in which sweeps are alternately made in the forward and backward directions of the ordering.[7] This is about as efficient as SOR. However, it has the advantage over SOR of having real eigenvalues and can be combined with *semi-iterative* methods (see Lecture 5) so as to achieve $O(h^{1/2})$ order of convergence with the 5-point approximation to the Dirichlet problem,[8] and D. M. Young has obtained a formula for the optimum overrelaxation factor (unpublished result).

Consistent ordering. In a similar vein, it has been shown that having a "consistent ordering" does not dramatically improve the rate of convergence [8],[9] and that all consistent orderings have exactly the same asymptotic rate of convergence.

7. Line and block overrelaxation.[10] The convergence of iterative methods can often be accelerated by using elimination to obtain a whole string of improved values at once. This is especially easy to achieve in the case of line or, more generally, k-line groupings of values on a rectangular network (2-cyclic case). The factor of acceleration for k-line overrelaxation is $\sqrt{k}$ [6], but not an order of magnitude (as $h \downarrow 0$). More important, such groupings make difference approximations *block tridiagonal* for sufficiently large k, permitting the use of block SOR.

In general, one can prove for (irreducible) Stieltjes matrices the much weaker result that block Gauss–Seidel converges more rapidly than point Gauss–Seidel; the proof involves "regular splittings" of matrices [V, p. 78]. More important, matrices arising from higher order problems such as the biharmonic equation of Example 3, §4, have block tridiagonal form relative to suitable k-line groupings of the mesh-lines. By applying block SOR to the resulting system, one can reduce the rate of convergence for the biharmonic equation from $O(h^4)$ to $O(h^2)$, for example.[11]

A much more intimate combination of partial elimination with iteration has been recently used by H. L. Stone and others[12] on problems arising from 5-point difference approximations to source problems. The basic idea is to construct a matrix B such that $A - B$ is readily factored, as $A - B = LU$, into lower (resp.

---

[7] J. W. Sheldon, Math. Tables Aid. Comput., 9 (1955), pp. 101–112; J. Assoc. Comput. Mach., 6 (1959), pp. 494–505.

[8] G. J. Habetler and E. L. Wachspress, Math. Comp., 15 (1961), pp. 356–362.

[9] See also C. G. Broyden, Numer. Math., 12 (1968), pp. 47–56.

[10] See [V, §6.4], [3], and the references given there. Early relevant papers include J. Schröder, Z. Angew Math. Mech., 34 (1954), pp. 241–253; R. J. Arms, L. D. Gates and B. Zondek, J. Soc. Indust. Appl. Math., 4 (1956), pp. 220–229; J. Heller, Ibid., 8 (1960), pp. 150–173.

[11] See S. V. Parter, Numer. Math., 1 (1959), pp. 240–252, and [6]; also J. Assoc. Comput. Mach., 8 (1961), pp. 359–365 and [V, p. 208].

[12] See [7]; also J. E. Gunn, SIAM J. Numer. Anal., 2 (1964), pp. 24–25; T. Dupont, Ibid., 4 (1968), pp. 753–782.

upper) triangular matrices L and U, and then to iterate

$$LU\mathbf{u}^{(r+1)} = B\mathbf{u}^{(r)} + \mathbf{k}.$$

Such "strongly implicit" iterative approaches deserve further study partly because of their potential adaptability to the variational formulations with piecewise polynomial approximations to be discussed in Lectures 7 and 8.

REFERENCES FOR LECTURE 4

[1] G. BIRKHOFF AND R. S. VARGA, *Reactor criticality and non-negative matrices*, J. Soc. Indust. Appl. Math., 6 (1958), pp. 354–377.
[2] G. H. GOLUB AND R. S. VARGA, *Chebyshev semi-iterative methods, successive overrelaxation iterative methods and second order Richardson iterative methods, I, II*, Numer. Math., 3 (1961), pp. 147–156; pp. 157–168.
[3] L. A. HAGEMAN AND R. S. VARGA, *Block iterative methods for cyclically reduced matrix equations*, Ibid., 6 (1964), pp. 106–119.
[4] L. A. HAGEMAN AND R. B. KELLOGG, *Estimating optimum overrelaxation parameters*, Math. Comp., 22 (1968), pp. 60–68.
[5] W. KAHAN, *Gauss–Seidel methods of solving large systems of linear equations*, Doctoral thesis, University of Toronto, 1958.
[6] S. V. PARTER, *Multi-line iterative methods for elliptic difference equations and fundamental frequencies*, Numer. Math., 3 (1961), pp. 305–319.
[7] H. L. STONE, *Iterative solution of implicit approximations of multidimensional partial differential equations*, SIAM J. Numer. Anal., 5 (1968), pp. 530–558. (See also T. Dupont, R. P. Kendall and H. H. Rachford, Ibid., pp. 559–573; and [BV, pp. 168–174].)
[8] R. S. VARGA, *Ordering of the successive overrelaxation scheme*, Pacific J. Math., 9 (1959), pp. 925–939.
[9] D. M. YOUNG, *Iterative methods for solving partial difference equations of elliptic type*, Trans. Amer. Math. Soc., 76 (1954), pp. 92–111.
[10] ———, *The numerical solution of elliptic and parabolic partial differential equations*, Modern Mathematics for the Engineer, Second Series, McGraw-Hill, New York, 1961, pp. 373–419.

LECTURE 5

Semi-iterative Methods

1. Chebyshev semi-iteration. In Lecture 4, I discussed purely iterative methods for solving $D\mathbf{u} = (E + F)\mathbf{u} + \mathbf{b}$, which can be reduced to $\mathbf{u} = B\mathbf{u} + \mathbf{k}$ with symmetrizable B by "scaling." These methods consist in applying repeatedly the linear inhomogeneous (affine) operator $L: \mathbf{u} \to B\mathbf{u} + \mathbf{k}$, i.e., in "iterating"

(1) $$\mathbf{u}^{(r)} = L[\mathbf{u}^{(r-1)}],$$

where (for example) we might have $L[\mathbf{u}] = B\mathbf{u} + \mathbf{k}$. In practice, optimal methods are seldom purely iterative, because numerical information obtained from previous iterations can usually be used as "feedback" to improve on L.

This leads to the study of *semi-iterative* methods of the more general form

(1') $$\mathbf{u}^{(r)} = L_r[\mathbf{u}^{(0)}, \mathbf{u}^{(1)}, \cdots, \mathbf{u}^{(r-1)}].$$

Specifically, we shall consider in this section the rate of convergence of methods of the form

(2) $$\mathbf{v}^{(r)} = \sum_{j=0}^{r} c_j^r \mathbf{u}^{(j)},$$

where $\mathbf{u}^{(j)} = B\mathbf{u}^{(j-1)} + \mathbf{k}$. To measure this, we define the *error* of an approximate solution $\mathbf{v}$ of $\mathbf{u} = B\mathbf{u} + \mathbf{k}$ as $\mathbf{e} = \mathbf{v} - \mathbf{u}$, where $\mathbf{u}$ is the exact solution of $\mathbf{u} = B\mathbf{u} + \mathbf{k}$. We shall consider only "solution preserving" methods such that $\mathbf{v}^{(0)} = \mathbf{u}$ implies $\mathbf{v}^{(r)} = \mathbf{u}$; hence $\mathbf{e}^{(r)} = \mathbf{v}^{(r)} - \mathbf{u} = 0$, for all $r > 0$. For this,

$$\sum_{j=0}^{r} c_j^r = 1$$

is necessary and sufficient. We shall then have

(3) $$\mathbf{e}^{(r)} = \sum_{j=0}^{r} c_j^r B^r[\mathbf{e}^{(0)}] = p_r(B)[\mathbf{e}^{(0)}], \quad p_r(x) = \sum_{j=0}^{r} c_j^r x^j.$$

For a general (random) initial $\mathbf{u}^{(0)} = \mathbf{v}^{(0)}$, the spectral radius of $p_r(B)$ provides the best measure of the rate of convergence. This leads one to ask: What choice of the c_j^r (i.e., of $p_r(B)$) will *minimize* the spectral radius of $p_r(B)$, among all polynomials p_r with $p_r(1) = 1$ (i.e., $\sum_{j=0}^{r} c_j^r = 1$)?

Since the eigenvalues of B are real and on $[-\rho(B), \rho(B)]$, the *Chebyshev polynomial*

(4) $$C_m(x/\beta)/C_m(\beta^{-1}), \quad \beta = \rho(B)$$

has the desired property. This was shown by L. F. Richardson, Lanczos [5], and Stiefel [8]. (Here $C_m(t) = \cos(m \cos^{-1} t)$ on $(-1, 1)$.) Moreover, from classic recursion formulas for the Chebyshev polynomials, it follows that[1]

$$\mathbf{u}^{(r)} = \omega_r \{B\mathbf{u}^{(r-1)} + \mathbf{k} - \mathbf{u}^{(r-2)}\} + \mathbf{u}^{(r-1)}, \qquad (5)$$

where the rth relaxation factor is

$$\omega_r = 1 + C_{r-2}(1/\rho)/C_r(1/\rho), \quad \rho = \rho(B).$$

Furthermore, as was first observed by Golub and Varga (Lecture 4, [2]),

$$\lim_{r \to \infty} \omega_r = \omega_b = 2/[1 + (1 - \rho^2(B))^{1/2}].$$

More generally, "for very large numbers of iterations, there is very little difference" between SOR and Chebyshev [V, p. 143], and Chebyshev "requires an additional vector of storage," which makes semi-iterative Chebyshev by itself asymptotically no better than SOR, as $r \to \infty$.

This additional storage can be eliminated by the semi-iterative *cyclic* Chebyshev methods which will now be described.[2] Recall that the semi-iterative method of (5) works whenever B is convergent ($\rho(B) < 1$) and Hermitian. When B is also weakly 2-*cyclic*, i.e., when

$$B = \begin{bmatrix} 0 & F \\ F^T & 0 \end{bmatrix}, \quad F \text{ Hermitian}, \qquad (5a)$$

we can partition the $\mathbf{u}^{(r)}$ of (5) as $\begin{bmatrix} \mathbf{u}_1^{(r)} \\ \mathbf{u}_2^{(r)} \end{bmatrix}$, corresponding to the splitting of B in (5a). Furthermore, taking appropriate components of the $\mathbf{u}^{(r)}$, the iteration of (5) reduces to

$$\begin{aligned} u_1^{(2m+1)} &= \omega_{2m+1}\{Fu_2^{(2m)} + k_1 - u_1^{(2m-1)}\} + u_1^{(2m-1)}, & m \geq 1, \\ u_2^{(2m+2)} &= \omega_{2m+2}\{F^T u_1^{(2m+1)} + k_2 - u_2^{(2m)}\} + u_2^{(2m)}, & m \geq 0, \end{aligned} \qquad (5b)$$

where ω_r is as before, with $\omega_1 = 1$, and where $u_1^{(1)} = Fu_2^{(0)} + k_1$. This semi-iterative cyclic Chebyshev method of Golub and Varga (Lecture 4, [2]) then requires no extra vector storage, and retains the superior norm characteristics of the cyclic Chebyshev method.

For very large systems of linear equations, combinations of multiline and block techniques with this semi-iterative cyclic Chebyshev method are probably the most effective methods in widespread use today. Here by multiline techniques, we mean direct inversion on sets of k adjacent lines.[3] Such multiline techniques permit one to adapt the cyclic Chebyshev method to the biharmonic and other higher order

---

[1] See A. Blair, N. Metropolis, et al., Math. Tables Aid. Comput., 13 (1959), pp. 145–184.
[2] The following exposition was kindly supplied by Professor Varga.
[3] E. Cuthill and R. S. Varga, J. Assoc. Comput. Mach., 6 (1959), pp. 236–244.

difference equations,[4] as well as to variational methods using piecewise polynomial approximations with patch bases (see Lecture 8).

2. Matrices H and V. A much more novel family of semi-iterative methods for solving plane elliptic problems is provided by *alternating direction implicit* (ADI) schemes, to which most of this lecture will be devoted.

As in Lecture 3, let the self-adjoint elliptic partial DE

(6) $$G(x, y)u - \frac{\partial}{\partial x}\left[A(x, y)\frac{\partial u}{\partial x}\right] - \frac{\partial}{\partial y}\left[C(x, y)\frac{\partial u}{\partial y}\right] = S(x, y)$$

be approximated by the 5-point difference equation

(7) $$(H + V + \Sigma)\mathbf{u} = \mathbf{b},$$

where, for a uniform rectangular mesh with mesh lengths h and k, we have

(8) $$Hu(x, y) = -a(x, y)u(x + h, y) + 2b(x, y)u(x, y) - c(x, y)u(x - h, y),$$

(9) $$Vu(x, y) = -\alpha(x, y)U(x, y + h) + 2\beta(x, y)u(x, y) - \gamma(x, y)u(x, y - k).$$

The most common choices for $a, b, c, \alpha, \beta, \gamma$ are

(10) $$\begin{aligned} a &= kA(x + h/2, y)/h, & c &= kA(x - h/2, y)/h, & 2b &= a + c, \\ \alpha &= hC(x, y + k/2)/k, & \gamma &= hC(x, y - k/2)/k, & 2\beta &= \alpha + \gamma. \end{aligned}$$

These choices[5] make H and V *symmetric* matrices acting on the vector space of functions $u = u(x_i, y_i)$ defined on interior mesh-points.

We shall assume that A and C are *positive* functions in (6) which makes the DE elliptic, while G is nonnegative. The matrix Σ is then a nonnegative diagonal matrix with diagonal entry $hkG(x_i, y_i)$ at (x_i, y_i). The vector $\mathbf{b}$ is computed by adding to the source terms $hkS(x_i, y_i)$ the terms in (8)–(9) associated with points on the *boundary* of the domain.

Our concern here is with the rapid solution of the vector equation (7) for large networks. For this purpose, it is essential to keep in mind some general properties of the matrices Σ, H and V.

As already stated, Σ is a nonnegative diagonal matrix. Moreover, H and V have positive diagonal entries and nonpositive off-diagonal entries. Because of the Dirichlet boundary conditions for (6), the diagonal dominance of H and V implies that they are positive definite [V, p. 23]; as in Lecture 4, such real symmetric and positive definite matrices with nonpositive off-diagonal entries are called *Stieltjes matrices*.

If the network $\mathscr{R}(h, h) = \mathscr{R}_h$ of interior mesh-points is connected, then $H + V$ and $H + V + \Sigma$ are also *irreducible*: it is known[6] that if a Stieltjes matrix is irreducible, then its matrix inverse has all positive entries.

---

[4] J. Heller, J. Soc. Indust. Appl. Math., 8 (1960), pp. 150–173; and S. V. Parter, Numer. Math., 3 (1961), pp. 305–319; see also Hageman and Varga (Lecture 4, [3]).

[5] See [1, § 2] for choices of a, b, c; also [W, pp. 70–74], Spanier [7, Part **d**] derives the appropriate difference approximations in a cylindrical or (r, z)-geometry.

[6] See [V, § 3.5]; irreducibility is defined in [V, § 1.4].

The matrices H and V are also *diagonally dominant*, by which we mean that the absolute value of the diagonal entry in any row is greater than or equal to the sum of the off-diagonal entries. For any $\theta \geqq 0$, the same is true a fortiori of $H + \theta\Sigma$, $V + \theta\Sigma$, and for $\theta_1 H + \theta_2 V + \theta\Sigma$ if $\theta_1 > 0$, $\theta_2 > 0$. The above matrices are all *diagonally dominant Stieltjes matrices*.

By ordering the mesh-points by rows, one can make H tridiagonal; by ordering them by columns, one can make V tridiagonal. That is, both H and V are similar to tridiagonal matrices, but one cannot make them both tridiagonal simultaneously.

3. Basic ADI operators. From now on, we shall consider only the iterative solution of the vector equation (7). Since it will no longer be necessary to distinguish the approximate solutions u from the exact solution $u(x, y)$, we shall cease to use boldface type, and will write u_n instead of $\mathbf{u}^{(n)}$.

Equation (7) is clearly equivalent, for any matrices D and E, to each of the two vector equations

(11) $\qquad (H + \Sigma + D)u = k - (V - D)u,$

(12) $\qquad (V + \Sigma + E)u = k - (H - E)u.$

This was first observed by Peaceman and Rachford in [6] for the case $\Sigma = 0$, $D = E = \rho I$ a scalar matrix. In this case, (11) and (12) reduce to

$$(H + \rho I)u = k - (V - \rho I)u, \quad (V + \rho I)u = k - (H - \rho I)v.$$

The generalization to $\Sigma \neq 0$ and arbitrary $D = E$ was made by Wachspress and Habetler [8].

For the case $\Sigma = 0$, $D = E = \rho I$ which they considered, Peaceman and Rachford proposed solving (7) by choosing an appropriate sequence of positive numbers ρ_n, and calculating the sequence of vectors $u_n, u_{n+1/2}$ defined from the sequence of matrices $D_n = E_n = \rho_n I$, by the formulas

(13) $\qquad (H + \Sigma + D_n)u_{n+1/2} = k - (V - D_n)u_n,$

(14) $\qquad (V + \Sigma + E_n)u_{n+1} = k - (H - E_n)u_{n+1/2}.$

Provided that $(H + \Sigma + D)$ and $(V + \Sigma + E)$ are nonsingular, and that the matrices to be inverted are similar under conjugation by permutation matrices (and scaling) to tridiagonal Stieltjes matrices, each of the equations (13) and (14) can be rapidly solved by Gauss elimination. The aim is to choose the initial trial vector u_0 and the matrices $D_1, E_1, D_2, E_2, \cdots$ so as to make the sequence $\{u_n\}$ converge rapidly.

Peaceman and Rachford considered the iteration of (13) and (14) when D_n and E_n are given by $D_n = \rho_n I$ and $E_n = \tilde{\rho}_n I$. This defines the Peaceman–Rachford method:

(15) $\qquad u_{n+1/2} = (H + \Sigma + \rho_n I)^{-1}[k - (V - \rho_n I)u_n],$

(16) $\qquad u_{n+1} = (V + \Sigma + \tilde{\rho}_n I)^{-1}[k - (H - \tilde{\rho}_n I)u_{n+1/2}].$

The rate of convergence will depend strongly on the choice of the iteration parameters $\rho_n, \tilde{\rho}_n$.

An interesting variant of the Peaceman–Rachford method was suggested by Douglas and Rachford [3, p. 422, (2.3)], again for the case $\Sigma = 0$. It can be defined for general $\Sigma \geqq 0$ by

$$u_{n+1/2} = (H_1 + \rho_n I)^{-1}[k - (V_1 - \rho_n I)u_n], \tag{17}$$

$$u_{n+1} = (V_1 + \rho_n I)^{-1}[V_1 u_n + \rho_n u_{n+1/2}], \tag{18}$$

where H_1 and V_1 are defined as $H + \frac{1}{2}\Sigma$ and $V + \frac{1}{2}\Sigma$, respectively. This amounts to setting $D_n = E_n = \rho_n I - \frac{1}{2}\Sigma$ in (13) and (14) and making some elementary manipulations. Hence (17) and (18) are also equivalent to (7) if $u_n = u_{n+1/2} = u_{n+1}$.

For higher-dimensional ADI methods, see J. Douglas, Numer. Math., 4 (1962), pp. 41–63, and J. Douglas, B. Kellogg and R. S. Varga, Math. Comp., 17 (1963), pp. 279–282.

4. Model problems. The power of ADI methods is greatest for model problems in which the preceding difference equations involve *permutable operators*, so that[7]

$$HV = VH, \quad H\Sigma = \Sigma H, \quad \text{and} \quad V\Sigma = \Sigma V. \tag{19}$$

This is the case if (1) reduces to the (modified) Helmholtz equation in a rectangle: $\sigma u - \nabla^2 u = S(x, y)$. More generally, H and V are permutable when the variables x and y are "separable" (in the sense discussed in Lecture 2, § 1) for the given elliptic problem.

If (19) holds, one can achieve an order-of-magnitude gain in the rate of convergence with the ADI methods described in § 3 by letting the ρ_n and $\tilde{\rho}_n$ be distributed in the intervals containing the (real) eigenvalues of $(H + \Sigma + \rho_n I)^{-1}(V - \rho_n I)$ and $(V + \Sigma + \tilde{\rho}_n I)^{-1}(H - \tilde{\rho}_n I)$ with equal proportionate spacing (see [7, g]). As the mesh-length h decreases, the number of (semi-) iterations required to reduce the error by a prescribed factor is (asymptotically and neglecting roundoff) only $O(\log h^{-1})$, as compared with $O(h^{-1})$ for SOR using the optimum relaxation parameter ω_b, or $O(h^{-2})$ as with Gauss–Seidel (or point-Jacobi).

A very interesting precise determination of the *optimum* parameters for such model problems has in fact been made by Jordan, in terms of elliptic functions; we shall omit the details.[8]

Unfortunately, it seems to be impossible to make rigorous extensions of the preceding theoretical results to most problems with variable coefficients or in nonrectangular regions [1], [2]. Whereas the theory of SOR applies to the 5-point approximation to general source problems, the experimentally observed success of ADI is in general hard to explain and even harder to predict.

---

[7] For discussions of (19), see [2, Part II], and R. E. Lynch, J. R. Rice and D. H. Thomas, Bull. Amer. Math. Soc., 70 (1964), pp. 378–384.

[8] See [W, p. 185], or E. Wachspress, J. Soc. Indust. Appl. Math., 10 (1962), pp. 339–350; 11 (1963), pp. 994–1016. Also C. de Boor and John Rice, Ibid., 11 (1963), pp. 159–169 and 12 (1964), pp. 892–896; and R. B. Kellogg and J. Spanier, Math. Comp., 19 (1965), pp. 448–452.

5. Iterative ADI. Even purely iterative (or "stationary") ADI methods using a single parameter ρ have the same order of convergence as optimized SOR [2, Theorem 20.1]. More generally [9, Theorem 1], simple iteration of the Peaceman–Rachford method (15)–(16) is always *convergent* if one chooses $D = E = \rho I - \Sigma/2$, where ρ is a positive number. This makes $D + \Sigma/2$ positive definite and symmetric and $H + V + \Sigma$ positive definite. A few sample proofs will be sketched below (see [1] and [2] for more details).

As in Lecture 4, we define the error vector as the difference $e_n = u_n - u_\infty$ between the *approximate solution* u_n after the nth iteration and the *exact solution* u_∞ of (7). For simplicity, we set $D = E = \rho I$. A straightforward calculation shows that, for the Peaceman–Rachford method, the effect of a single iteration of (15)–(16) is to multiply the error vector e_n by the error reduction matrix T, defined by

$$(20) \qquad T_\rho = (V + \Sigma + \rho I)^{-1}(H - \rho I)(H + \Sigma + \rho I)^{-1}(V - \rho I).$$

Likewise, the error reduction matrix for the Douglas–Rachford method (18)–(19) with all $\rho_n = \rho$ is given by

$$(21) \qquad \begin{aligned} W_\rho &= (V_1 + \rho I)^{-1}(H_1 + \rho I)^{-1}(H_1 V_1 + \rho^2 I) \\ &= [H_1 V_1 + \rho(V_1 + H_1) + \rho^2 I]^{-1}(H_1 V_1 + \rho^2 I). \end{aligned}$$

If one assumes that $D_n = -\Sigma/2 + \rho I = E_n$ also for the generalized Peaceman–Rachford method (13)–(14), then from (15), we have

$$(22) \qquad T_\rho = (V_1 + \rho I)^{-1}(H_1 - \rho I)(H_1 + \rho I)^{-1}(V_1 - \rho I),$$

and the matrices W_ρ and T_ρ are related by

$$(23) \qquad 2W_\rho = I + T_\rho.$$

We next prove a lemma which expresses the algebraic content of a theorem of Wachspress and Habetler [9, Theorem 1].

LEMMA 1. *Let P and S be positive definite real matrices, with S symmetric. Then $Q = (P - S)(P + S)^{-1}$ is norm-reducing[9] for real row vectors x relative to the norm $\|x\| = (xS^{-1}x^T)^{1/2}$.*

Proof. For any norm $\|x\|$, the statement that Q is norm-reducing is equivalent to the statement that $\|(S - P)y\|^2 < \|(S + P)y\|^2$ for every nonzero vector $y = (P + S)^{-1}x$. In turn, this is equivalent for the special Euclidean norm $\|x\| = (xS^{-1}x^T)^{1/2}$ to the statement that

$$(24) \qquad y(P + S)S^{-1}(P^T + S^T)y^T > y(P - S)S^{-1}(P - S)^T y^T$$

for all nonzero y. Expanding the bilinear terms, cancelling, and dividing by two, this is equivalent to the condition that $y(P + P^T)y^T > 0$ for all nonzero y. But this is the hypothesis that P is positive definite.[10]

COROLLARY. *In Lemma 1, $\rho(Q) < 1$.*

---

[9] The phase "norm-reducing" here refers to Euclidean norm only in special cases.

[10] Note that P is *not* assumed to be symmetric, but only to be such that $x^T(P + P^T)x > 0$, for all real $x \neq 0$.

This follows from Lemma 1 and the following general result on matrices:

$$\rho(M) \leq \max_{\|x\|=1} (\|Mx\|/\|x\|) \quad \text{for any norm } \|\cdot\|.$$

Actually, $\rho(M)$ is the infimum of $\max(\|Mx\|/\|x\|)$ taken over all Euclidean (inner product) norms.

THEOREM 1. *Any iterative ADI process* (13)–(14) *with all* $D_n = D$ *and all* $E_n = E$ *is convergent, provided* $\Sigma + D + E$ *is symmetric and positive definite, and* $2H + \Sigma + D - E$ *and* $2V + \Sigma + E - D$ *are positive definite.*[11]

Proof. It suffices to show that $\rho(T) < 1$. But since similar matrices have the same eigenvalues and hence the same spectral radius, the error reduction matrix

(25) $\qquad T = (V + \Sigma + E)^{-1}(H - E)(H + \Sigma + D)^{-1}(V - D)$

of (13)–(14) has the same spectral radius as

(26) $\qquad \begin{aligned} \tilde{T} &= (V + \Sigma + E)T(V + \Sigma + E)^{-1} \\ &= [(H - E)(H + \Sigma + D)^{-1}][(V - D)(V + \Sigma + D)^{-1}]. \end{aligned}$

By Lemma 2, both factors in square brackets reduce the norm $[x^T(\Sigma + D + E)^{-1}x]^{1/2} = \|x\|$, provided $\Sigma + D + E = 2S$, $R_H = [H + \Sigma/2 + (D - E)/2]$ and $R_V = [V + \Sigma/2 + (E - D)/2]$ are positive definite, and $\Sigma + D + E$ is also symmetric.

It is easy to apply the preceding result to difference equations (8)–(9) arising from the Dirichlet problem for the self-adjoint elliptic differential equation (6). In this case, as stated in § 2, H and V are diagonally dominant (positive definite) *Stieltjes matrices*. The same properties hold a fortiori for $\theta_1 H + \theta_2 V + \theta_3 \Sigma$ if all $\theta_i \geq 0$ and $\theta_1 + \theta_2 > 0$.

Hence the hypotheses of Theorem 1 are fulfilled for $D = \rho I - \theta \Sigma, E = \tilde{\rho} I - \tilde{\theta} \Sigma$ for any $\rho, \tilde{\rho} > 0$ and any $\theta, \tilde{\theta}$ with $0 \leq \theta, \tilde{\theta} \leq 2$. Substituting into (13)–(14), we obtain the following result.

COROLLARY 1. *If* $\rho, \tilde{\rho}, > 0$ *and* $0 \leq \theta, \tilde{\theta} \leq 2$, *then the stationary ADI method defined with* $\theta' = 2 - \theta$ *by*

(27) $\qquad (H + \theta\Sigma/2 + \rho I)u_{n+1/2} = k - (V + \theta'\Sigma/2 - \rho I)u_n,$

(28) $\qquad (V + \theta\Sigma/2 + \rho I)u_{n+1} = k - (H + \theta'\Sigma/2 - \rho I)u_{n+1/2}$

is convergent.

In fact, it is norm-reducing for the norm defined by

$$\|x\|^2 = x^T(\Sigma + D + E)^{-1}x = x^T[(\rho + \tilde{\rho})I + (\theta + \tilde{\theta})\Sigma/2]^{-1}x.$$

6. Final remarks. One of the most interesting treatments of a reasonably general case is that of Guilinger [4]. Utilizing the *smoothness* of solutions of elliptic DE's (see Lecture 2), Guilinger proved that the Peaceman–Rachford semi-iterative method could be made to reduce the error by a given factor in a number of steps which was *independent of the mesh-length*.[12]

---

[11] This result, for $D - E = 0$, was first given in [9]. For the analogous result on W, see [1].
[12] See also R. E. Lynch and J. R. Rice, Math. Comp., 22 (1968), pp. 311–335.

Also, Widlund [10] has obtained some theoretical results for ADI with variable ρ (i.e., as a semi-iterative method) in the noncommutative case. Moreover, Spanier [7] and Kellogg[13] have applied ADI methods to ΔE's on nonrectangular meshes.

Finally, the reader's attention is called to the existence of a carefully documented[14] HOT-1 code, written at the Bettis Atomic Power Laboratory, whose relation to the theoretical principles described in this chapter has been the subject of a careful and lucid exposition by Spanier [7].

REFERENCES FOR LECTURE 5

[1] G. BIRKHOFF AND R. S. VARGA, *Implicit alternating direction methods*, Trans. Amer. Math. Soc., 92 (1959), pp. 13–24.

[2] G. BIRKHOFF, R. S. VARGA AND DAVID YOUNG, *Alternating direction implicit methods*, Advances in Computers, 3 (1962), pp. 189–273.

[3] J. DOUGLAS, JR. AND H. RACHFORD, *On the numerical solution of heat conduction problems in two and three space variables*, Trans. Amer. Math. Soc., 82 (1956), pp. 421–439.

[4] W. H. GUILINGER, *Peaceman-Rachford method with small mesh-increments*, J. Math. Anal. Appl., 11 (1964), pp. 261–277.

[5] C. LANCZOS, *Solution of systems of linear equations by minimized iterations*, J. Res. Nat. Bur. Standards, 49 (1952), pp. 33–53.

[6] D. W. PEACEMAN AND H. H. RACHFORD, JR., *The numerical solution of parabolic and elliptic differential equations*, J. Soc. Indust. Appl. Math., 3 (1955), pp. 28–41.

[7] J. SPANIER, *Alternating direction methods applied to heat conduction problems*, Mathematical Methods for Digital Computers, A. Ralston and H. S. Wilf, eds., vol. II, John Wiley, New York, 1967, pp. 215–245.

[8] E. STIEFEL, *On solving Fredholm integral equations*, J. Soc. Indust. Appl. Math., 4 (1956), pp. 63–85.

[9] E. L. WACHSPRESS AND G. J. HABETLER, *An alternating-direction-implicit iteration technique*, Ibid., 8 (1960), pp. 403–424.

[10] O. WIDLUND, *On the rate of convergence of an alternating direction implicit method*, Math. Comp., 20 (1966), pp. 500–515.

[11] DAVID YOUNG, *On the solution of linear systems by iteration*, Proc. Symposia Applied Math., vol. VI, American Mathematical Society, Providence, 1956, pp. 283–298.

---

[13] R. B. Kellogg, Math. Comp., 18 (1964), pp. 203–210.

[14] R. B. Smith and J. Spanier, Bettis Atomic Power Laboratory Report WAPD-TM-465, 1964.

LECTURE 6

Integral Equation Methods

1. Introduction. The last three lectures were devoted to difference methods. These quickly reduce elliptic problems to an approximately equivalent algebraic form by elementary considerations from analysis; the main job is to solve the resulting system of algebraic equations.

The next three lectures will make much deeper use of classical analysis, including especially more sophisticated *approximation* methods (in Lecture 7) and *variational* methods (in Lecture 8). The present lecture will introduce the subject by describing briefly a number of numerical techniques which are especially closely related to the results from classical analysis reviewed in Lecture 2.

These techniques are especially applicable to homogeneous linear elliptic DE's with constant coefficients, such as $\nabla^2 u = 0$ or $\nabla^4 u = 0$. One of the most powerful classical techniques consists in expanding in series. We already saw in Lecture 2 how effective this technique was for solving the Dirichlet problem in the unit disc (by Fourier series). In §2, we shall discuss its extension on to other domains.

A related but more sophisticated approach consists in expressing functions in terms of definite integrals of their boundary values or other quantities (e.g., their normal derivatives on the boundary). This approach will be discussed in §3 and §4.

Both techniques rely essentially on the principle that any (discrete or continuous) *superposition* (linear combination or integral) of solutions of a given homogeneous linear DE is again a solution. Hence, if one has a *basis* of elementary solutions, one can take the coefficients w_j or weight-function $w(\mathbf{s})$ as unknowns in an expression for the general solution

$$u(\mathbf{x}) = \Sigma w_j \varphi_y(\mathbf{x}) \quad \text{or} \quad u(\mathbf{x}) = \int \varphi(\mathbf{s}, \mathbf{x}) \, dw(\mathbf{s}),$$

respectively, and then try to obtain enough equations on the w_j or on $w(\mathbf{s})$ to determine which of them represents the solution.

This lecture and Lecture 8 will contain several illustrations of ways to obtain numerical results by implementing the above principle. In general, one must use (approximate) numerical quadrature to obtain such results, although in exceptional cases formal integration may be possible.

2. Superposition of elementary solutions. By the maximum principle, any harmonic function $u(\mathbf{x})$ which is uniformly approximated on the boundary ∂R of a compact region R by a harmonic polynomial $h(\mathbf{x})$ will also be uniformly approximated by $h(\mathbf{x})$ in the interior of R. On the other hand, classic theorems of Bergman and Walsh assert that every harmonic function in a compact simply connected domain can be so approximated. These theorems, in turn, generalize an earlier theorem of Runge[1] which asserts that any complex analytic function in R can be uniformly approximated by a complex polynomial, a result which generalizes the classic Weierstrass approximation theorem (for which see Lecture 7, § 3).

Therefore, given $f(\mathbf{x})$ on the boundary ∂R of a plane region R, one can look for the harmonic polynomial $h_n(\mathbf{x})$ of a degree n which gives the best mean square (or a best uniform) approximation to $f(\mathbf{x})$ on ∂R, and use $h_n(\mathbf{x})$ to approximate the solution of the Dirichlet problem in R for the specified boundary conditions.

This approach has been used successfully by Bergman and others to solve the Dirichlet and other similar elliptic boundary value problems. Bergman and Herriot describe the method as follows:[2]

The method of the kernel function for solving elliptic boundary value problems consists essentially of three steps:
1. A procedure for generating a (complete) set (h_ν), $\nu = 1, 2, \cdots$, of particular solutions of the partial differential equation.
2. A procedure for deriving from this system a set of particular solutions (ψ_ν), $\nu = 1, 2, \cdots$, which are orthonormal in a suitable sense over a given domain.
3. A procedure for determining from the prescribed boundary values a linear combination $\sum_{\nu=1}^{N} a_\nu \psi_\nu$ approximating the desired solution.

Bergman and Herriot used this method to solve a Dirichlet problem for the DE

$$\nabla^2 u + (2r^2 - 4)u = 0, \qquad r^2 = x^2 + y^2,$$

in the square $|x| \leq 1$, $|y| \leq 1$ with circularly rounded corners, for boundary values having the 8-folded symmetry of the domain. One can greatly reduce roundoff error accumulation (ill-conditioned matrices) by orthonormalizing the basis of polynomial solutions. Techniques for doing this are described by Davis and Rabinowitz,[3] who have applied the method successfully to solve the Dirichlet problem for $\nabla^2 u = 0$ in a cube, a sphere, an elliptic annulus, and a square with the corners cut out, as well as for $\nabla^4 u = 0$ in various domains.

An essential feature of this and the other methods to be discussed in this lecture lies in their use of integral identities to characterize functions of $n = 2$ or 3 variables satisfying some differential equation in terms of a function of one less variable on the boundary, which is usually easier to compute with and to visualize (through its graph). Such methods are especially well-adapted to solving the Laplace and related equations for which Green's identities are relatively simple. They are not

---

[1] C. Runge, Acta Math., 6 (1885), pp. 229–244; S. Bergman, Math. Ann., 86 (1922), pp. 238–271; Lecture 7, [15], esp. pp. 19, 35.

[2] Numer. Math., 7 (1965), pp. 42–65. Cf. also Ibid., 3 (1961), pp. 209–225; K. T. Hahn, Pacific J. Math., 14 (1964), pp. 944–955; S. Bergman and B. Chalmers, Math. Comp., 21 (1967), pp. 527–542.

[3] Advances in Computers, 2 (1961), pp. 55–133; see also J. Assoc. Comput. Mach., 1 (1954), pp. 183–191.

well-adapted to nonlinear DE's or to DE's with variable coefficients; hence they are much less generally applicable than difference methods.

3. L-shaped membrane. In general, a self-adjoint elliptic differential operator on a compact domain has an infinite sequence of real *eigenvalues* λ_n, such that

$$L[\varphi_n] + \lambda_n \varphi_n = 0,$$

and associated eigenfunctions φ_n. Moreover, these eigenfunctions are complete, a fact which was guessed by the physicists Ohm and Rayleigh 50 years before it was proved rigorously (Ohm–Rayleigh principle). The computation of the sequence of eigenvalues for a specified region is a standard problem of analysis, for whose approximate solution recourse must be had to numerical methods in most cases (cf. Lecture 2, § 1). My article with George Fix in [BV] gives a careful review of the general problem.

One such eigenvalue problem, on which much computational effort has been expended, is the determination of the eigenvalues of an L-shaped membrane. By Lehman's theorem (Lecture 2, § 7), we know that the corresponding eigenfunctions are analytic except at the reentrant corner, where they are asymptotic to

(1) $$J_{2n/3}(k_n r) \sin(2n\pi\theta/3), \qquad k_n^2 = \lambda,$$

or to a linear combination of such functions.

This fact led Fox, Henrici and Moler[4] to apply the superposition method of § 2 to the preceding problem, taking as basis functions of the prescribed form (1). For such applications to eigenvalue problems, each eigenvalue must be taken as a parameter to be determined by successive approximation.

Using this procedure, extremely accurate eigenvalues for the L-shaped membrane were obtained in the paper cited. Only very recently, using the approximation and Rayleigh–Ritz techniques to be described in Lectures 7 and 8, have Fix and Wakoff obtained more accurate results.[5]

4. Green's identities. One of the most familiar formulas of complex analysis is Cauchy's integral formula

(2) $$f(z) = \frac{1}{2\pi i} \oint_C \frac{f(t)\, dt}{(t-z)}.$$

This enables one to calculate the values of any complex analytic function in the interior of a closed curve C in the complex plane from its values on the boundary, and is a cornerstone for the general theory.

Actually, writing

(3) $$w = f(z) = f(x+iy) = u(x,y) + iv(x,y),$$

---

[4] SIAM J. Numer. Anal., 4 (1967), pp. 89–102.
[5] G. J. Fix and G. I. Wakoff, J. Comp. Phys., to appear.

either u or its conjugate v on C determines the interior values of both up to an additive constant. Thus, if C is the unit circle, $z = e^{i\theta}$, then on C the formula

$$(4) \qquad v(\theta) = v_0 + \frac{1}{2\pi}\oint \cot[(\theta - \sigma)/2]u(\sigma)\,d\sigma$$

expresses v in terms of u. Formula (4) is valid if the Cauchy principal value is taken and f is regular in $|z| < 1$ and continuous in $|z| \leq 1$. One can easily derive the "singular convolution kernel" $\cot[(\theta - \sigma)/2]$ in (4) formally, using Fourier series. These give

$$v(\theta) = \frac{1}{\pi}\sum_{k=1}^{\infty}\left[\int \cos k\sigma \sin k\theta + \sin k\sigma \cos k\theta\right]u(\sigma)\,d\sigma + v_0.$$

Setting $\varphi = \sigma - \theta$ and $\sum_{k=1}^{\infty} \cos k\varphi = \frac{1}{2}\cot \varphi/2$, we obtain (4).

Closely related to Cauchy's integral formula is Green's third identity:

$$(5) \qquad u(x_0, y_0) = \frac{1}{2\pi}\oint_C \left\{\frac{\partial}{\partial v}(\log r)u - (\log r)\frac{\partial u}{\partial v}\right\}ds,$$

where $r^2 = [(x - x_0)^2 + (y - y_0)]^2$. This formula is valid at any interior point x_0, y_0 of *any* bounded domain bounded by a smooth curve C, for any function $u(x, y)$ which is harmonic interior to C. In three dimensions, Green's third identity is, similarly,

$$(6) \qquad u(x_0, y_0, z_0) = -\frac{1}{4\pi}\int\int_S \left\{\frac{\partial}{\partial v}\left(\frac{1}{r}\right)u - \frac{1}{r}\frac{\partial u}{\partial v}\right\}dS,$$

valid for any function $u(x, y, z)$ which is harmonic in the interior R of a bounding surface $S = \partial R$.

Instead of the "fundamental solutions" $(\log r)/2\pi$ and $r/4\pi$ of the Laplace equation in two and three dimensions, respectively, we now use the *Green's functions* $G(x, y; x_0, y_0)$ and $G(x, y, z; x_0, y_0, z_0)$ for the Dirichlet problem in R. By definition, these are harmonic functions which (i) vanish on ∂R, and (ii) differ from the fundamental solutions near the "poles" (x_0, y_0) (resp. (x_0, y_0, z_0)) by bounded quantities. The argument leading to (5) and (6) then gives as in Lecture 2, (7),

$$(7) \qquad u(\mathbf{x}) = \int_{\partial R}\frac{\partial G}{\partial v}(\mathbf{x};\xi)u(\xi)\,d\xi.$$

In the particular case of a sphere, therefore, $\partial G/\partial v$ is the kernel function of the Poisson integral formula for the sphere. Unfortunately, the kernel $\partial G(\mathbf{x}, \xi)/\partial v$ depends on five variables in three space dimensions; hence (7) is not practical computationally in most spatial domains.

5. Integral equations of potential theory. Green's third identity can also be interpreted as asserting that (under suitable differentiability hypotheses) every function harmonic on the closure $\bar R = R \cup \partial R$ of R can be expressed as the potential of a distribution or "spread" on ∂R of poles *and* dipoles normal to ∂R. However, much as a function u harmonic in a compact region R is determined by either its values u *or* its normal derivatives on R, so u can be represented as the potential of a spread of poles on ∂R (a "single layer"), *or* as the potential of a spread of normal dipoles (a "double layer"). If ∂R is smooth, moreover, the *densities* $\rho(x)$, $\sigma(x)$ of these spreads vary smoothly. Kantorovich and [KK, pp. 130–140] describe a numerical technique of Krylov and Bogoliubov (1929) for implementing this numerically to solve the Dirichlet problem.[6]

Finally, if one takes these densities as unknown functions, then the condition that the resulting potential have given values $u = f(\mathbf{x})$ (resp. normal derivatives $g(\mathbf{x}) = \partial u/\partial v$ on ∂R) can be shown to be equivalent to *Fredholm integral* equations whose forms are

$$(8) \qquad f(\mathbf{x}) = \sigma(\mathbf{x}) - \int K(\mathbf{x}, \xi)\sigma(\xi)\,d\xi,$$

(resp.

$$(9) \qquad g(\mathbf{x}) = \rho(\mathbf{x}) + \int K(\xi, \mathbf{x})\rho(\xi)\,d\xi\,)$$

for a suitable kernel $K(\mathbf{x}, \xi)$, which is $(1/2\pi)(\partial/\partial v)(1/r)$ in three dimensions.

The integral equations (8) and (9) are sometimes called "the integral equations of potential theory"; they have received an enormous amount of study. For their derivation, see [K, pp. 286–291] and Garabedian (Lecture 2, [6, §9.3]). By the "Fredholm alternative," uniqueness implies existence.

6. Conformal mapping; Gerschgorin's method. Riemann's mapping theorem asserts the following: *If R is any simply connected region of the complex ζ-plane with smooth boundary ∂R, and ζ_0, ζ_1, are points inside R and on ∂R, respectively, then there exists a unique one-one conformal map of R onto the unit disc $r < 1$, which maps ζ_0 into 0 and ζ_1 into 1.*

The truth of this theorem follows from the existence of a Green's function in R for the "pole" ζ_0. Set $\log r = -G(\zeta, \zeta_0)$, and θ as the conjugate function

$$(10) \qquad \theta = \int_{\zeta_1}^{\zeta} \left[\frac{\partial G}{\partial y}dx - \frac{\partial G}{\partial x}dy \right].$$

Then the mapping $\zeta \to re^{i\varphi}$ (polar coordinates) achieves the desired result.

Likewise, there exists a conformal map from any *doubly* connected plane region onto the annulus $1 < r < \gamma$ for some $\gamma > 1$. This theorem is plausible from the following physical interpretation. Consider the equilibrium temperature $\tau(x, y)$ for

---

[6] See also M. S. Lynn and W. R. Timlake, Numer. Math., 11 (1968), pp. 77–98.

boundary values 0 on the inner boundary and τ_1 on the outer boundary. Again let θ be the conjugate function, and choose τ_1 so that

$$-\oint \left(\frac{\partial \tau}{\partial y} dx - \frac{\partial \tau}{\partial x}\right) dy = 2\pi.$$

By the strict maximum principle, τ can have no critical points in the domain; hence for $\tau = \log r$ and $\tau_1 = \log \gamma$, the mapping $\zeta \to re^{i\varphi}$ should achieve the desired result.

Many numerical techniques for conformal mapping are described in [KK, Chap. VI], and in Gaier [3]. We shall here present first Gerschgorin's method, following especially [2, Theorem 1] (see also [KK, § V, 9]). To this end, we suppose ∂R given parametrically by $z = z(q)$, where q is a periodic variable with period 2π, and set

(11) $$A(z_j, z) = \frac{1}{\pi \rho_j} \frac{\partial \rho_j}{\partial n} \frac{ds}{dq}, \quad \rho_j = |z - z_j|, \quad ds = |dz|.$$

The solution is then given by the "angular distortion" $u(z)$, which satisfies

(12) $$u(z_j) = \lambda \oint A(z_j, z) u(z)\, dq + \Phi(z_j),$$

where $\lambda = +1$ for interior and $\lambda = -1$ for exterior mappings, and

(12') $$\Phi(z_j) = \pi^{-1} \oint (\log r)\, d\rho_j/\rho_j.$$

Setting $z_j = z(q_j) = z(2\pi j/n)$, and using trapezoidal quadrature for maximum accuracy [2, § 4], we can compute approximate values $u_n(z_j)$ of the $u(z_j)$. As $n \to \infty$, the $u_n(z)$ as computed from (12) converge to the desired map.

If C is any strictly *convex* curve, then $A(z_j; z)$ is positive. Hence, by Jentzsch's theorem, simple iteration will converge (see also [7] and, for early numerical experiments, [9]).[7]

7. Theodorsen's method. For applications to (two-dimensional) airfoil theory (see Lecture 9, § 2), one wants to map the exterior of the unit circle $\Gamma: r = e^{i\sigma}$ conformally onto the exterior of some specified curve γ. For such applications, another (formerly popular) numerical method of conformal mapping is that of Theodorsen [6][8]; it may be described as follows [2, § 6].

Let γ be specified by $r = e^{F(\theta)}$. For any periodic function $g(\sigma)$, the conjugate is

(13) $$C[g(\sigma)] = \frac{1}{2\pi} \oint \cot \frac{\sigma - \sigma'}{2} g(\sigma')\, d\sigma',$$

---

[7] For applications of the Gerschgorin method to domain perturbation, see [3, pp. 59–60] and S. Warschawski, J. Math. Mech., 19 (1970), pp. 1131–1153. Even if R has corners, one can adapt Gerschgorin's method; see [3, pp. 16–21].

[8] See also T. Theodorsen and I. E. Garrick, NACA Tech. rep. 452, 1934.

as in (4). Take as an unknown function the angular distortion $u(\sigma) = \sigma - \theta$ with conjugate $v = \log r$, for the desired conformal map of $re^{i\sigma}$ onto γ. Since $v = -F(\sigma - u(\sigma))$ on γ, we clearly have

$$(13') \qquad u(\sigma) = \pm C[F(\sigma - u(\sigma))]$$

with the plus sign for the exterior mapping and the minus sign for the interior mapping. The nonlinear integral equation (13)–(13') can also often be solved by simple iteration.[9]

Analytic domain perturbation. Theodorsen's method is most effective when γ is a small perturbation of the unit circle Γ. In the case of *analytic* domain perturbation, it is worth trying a very different *perturbed power series* technique due to Kantorovich.[10] For example, consider the family of coaxial ellipses

$$(14) \qquad x^2 + y^2 - 2\lambda(x^2 - y^2) = 1.$$

For each λ, the appropriate mapping is given by

$$(14') \qquad z = f(\zeta; \lambda) = \sum_{k=1}^{\infty} a_k(\lambda)\zeta^k,$$

where power series in λ for the $a_k(\lambda)$ can be computed recursively. Up to an error of $O(\lambda^6)$, one finds

$$(15) \qquad \begin{aligned} z = (1 - \lambda^2/2 + 3\lambda^4/8)[\zeta + (\lambda - 2\lambda^3 + 3\lambda^5)\zeta^3 \\ + (2\lambda^2 - 9\lambda^4)\zeta^5 + (5\lambda^3 - 36\lambda^5)\zeta^7 + 14\lambda^4\zeta^9 + 42\zeta^{11}]. \end{aligned}$$

Other methods of conformal mapping will be discussed in Lecture 8. For a very complete review of the subject, which covers multiply connected regions and lists 480 references for further study, consult Gaier [3]. Note also the availability of a very complete dictionary of special conformal transformations.[11]

REFERENCES FOR LECTURE 6

[1] S. BERGMAN AND J. G. HERRIOT, *Application of the method of the kernel function for solving boundary-value problems*, Numer. Math., 3 (1961), pp. 209–225.
[2] G. BIRKHOFF, D. M. YOUNG AND E. H. ZARANTONELLO, *Numerical methods in conformal mapping*, Proc. Symposia Applied Math., vol. IV, American Mathematical Society, Providence, 1953, pp. 117–140.
[3] D. GAIER, *Konstruktive Methoden der konformen Abbildung*, Springer, Berlin, 1964.
[4] S. GERSCHGORIN, *On the conformal mapping of a simply connected region onto a circle*, Mat. Sbornik, 40 (1933), pp. 48–58. (In Russian.)
[5] N. I. MUSKHELISHVILI, *Some Basic Problems in the Mathematical Theory of Elasticity*, Noordhoff, Groningen, 1953.

---

[9] For a general discussion of numerical methods for solving nonlinear integral equations, see Ben Noble's article in *Nonlinear Integral Equations*, P. M. Anselone, editor, University of Wisconsin Press, Madison, 1964, pp. 215–317.

[10] [KK, Chap. V, §6] and [3, pp. 166–168]. For the original presentation, see L. V. Kantorovich, Mat. Sbornik, 40 (1933), pp. 294–325.

[11] H. Kober, *Dictionary of Conformal Representations*, Dover, New York, 1952.

[6] T. THEODORSEN, *Theory of wing sections of arbitrary shape*, NACA Tech. rep. 411, 1931.
[7] S. WARSCHAWSKI, *Recent results in numerical methods in conformal mapping*, Proc. Symposia Applied Math., vol. VI, American Mathematical Society, Providence, 1956, pp. 219–250.
[8] *Construction and application of conformal maps*, Nat. Bureau Standards Appl. Math. Series 18, 1952.
[9] *Experiments in the computation of conformal maps*, Nat. Bureau Standards Appl. Math. Series 42, 1955, esp. pp. 31–44.

See also P. Davis and P. Rabinowitz, Advances in Computers, vol. 2, on orthonormalization.

LECTURE 7

Approximation of Smooth Functions

1. Classical univariate interpolation. There is an enormous classical literature concerned with approximations to functions of *one* real or complex variable (see [4], [8], [9], [10], [14], [15]). The problem of finding effective approximation methods for functions of two or more variables, which is my main concern here, is very different and, in general, much harder. Therefore, I shall here only recall a few essential results from the classical literature about univariate approximation. In his Regional Conference Lectures, Professor Varga will give a much fuller coverage.[1]

The most widely used approximation formulas are based on algebraic interpolation schemes, and are therefore exact (for the function, though not for its derivatives) at mesh-points. The simplest of these is Lagrange interpolation. It is classic that, given $x_0 < \cdots < x_n$ and $f(x_i) = y_i$, $i = 0, 1, \cdots, n$, there is one and only one polynomial $p(x)$ of degree n such that $p(x_i) = y_i$. Moreover, the *interpolation error* is given on $[x_0, x_n]$ by

$$(1) \qquad f(x) - p_n(x) = e_n(x) = \prod_{i=0}^{n} (x - x_i) f^{(n+1)}(\xi)/(n+1)!, \quad \xi \in (x_0, x_n).$$

Setting $x_i = x_0 + h\theta_i$, $i = 0, \cdots, n$, we therefore have

$$(2) \qquad |e_n(x)| \leq [h^{n+1}/(n+1)!] \prod_{i=1}^{n} |\zeta - \theta_i| \cdot |f^{(n+1)}(\xi)| = O(h^{n+1}),$$

where $0 < \zeta = (x - x_0)/(x_n - x_0) < 1$ and $\xi \in (x_0, x_n)$. More generally, for any function $f \in C^{n+1}[a, b]$, Lagrange polynomial interpolation of degree n gives a jth derivative whose errors are $O(h^{n+1-j})$ for $j \in 0, 1, \cdots, n$, as $h \downarrow 0$ with fixed n.

In the limiting case of coincident points, we get Hermite interpolation. For example, the *cubic Hermite* interpolant through $x_0 = x_1 = a$, $x_2 = x_3 = b$ is that cubic polynomial such that

$$p(a) = y_0, \quad p'(a) = y'_0, \quad p(b) = y_1, \quad p'(b) = y'_1.$$

More generally the *Hermite interpolant* of degree $2m - 1$ to $f(x)$ on $[a, b]$ is that polynomial $p(x) = c_0 + c_1 x + \cdots + c_{0n}x^n$, $n = 2m - 1$, such that $p^{(k)}(a) = u^{(k)}(a)$ and $p^{(k)}(b) = u^{(k)}(b)$ for $k = 0, \cdots, m - 1$ [4, p. 28]. For functions of class $C^{n+1}[a, b]$,

---

[1] Those wishing to get the flavor of contemporary "pure" research in this area might also peruse A. Talbot, *Approximation Theory*, Academic Press, New York, 1970.

Hermite interpolation of fixed degree n also approximates the jth derivative with error $O(h^{n+1-j})$, $j = 0, 1, \cdots, n$, as $h \downarrow 0$.

On a fixed interval $[a, b]$, the convergence as $n \to \infty$ of a sequence of Lagrange interpolants $p_n(x)$ of degree n to a given function $f(x)$ depends in general on the location of the mesh-points. The use of a uniform mesh with $x_i = a + (b - a)i/n$ can give very poor results.

Thus, the Lagrange interpolants $p_n(x)$ of degree n to a given function $f(x)$ on a fixed interval $[a, b]$, over a uniform mesh with $x_i = a + i(b - a)/n$, need not even converge to that function. Thus, as a famous counterexample of Runge shows [7, Chap. V, §15], one need not have $\lim_{n \to \infty} p_n(x) \to f(x)$ even for a bounded analytic function like $1/(1 + x^2)$ on $[-\sqrt{5}, \sqrt{5}]$. This is essentially because of the imaginary poles of $1/(1 + x^2)$ at $x = \pm i$.

Analytic periodic functions. The situation is more favorable for smooth periodic functions $f(\theta)$, with $f(\theta + 2\pi) = f(\theta)$. Here interpolation by trigonometric polynomials of degree n through $2n$ or $2n + 1$ equidistant points converges exponentially to the function. For $f \in C^{(n)}$, with $f^{(k)} \in \text{Lip } \alpha$, the error is $O(n^{-k-\alpha})$.

Functions analytic on $[-1, 1]$. Similar results hold for real functions $g(x)$ analytic on a real interval, which we can translate and scale to be $[-1, 1]$. Since $[-1, 1]$ is compact, $g(x)$ can be extended to a function $g(z)$ analytic in some ellipse with foci at ± 1. Setting $z = (t + t^{-1})/2$, we obtain a function $f(t)$ analytic on an annulus $-\varepsilon < \log|t| < \varepsilon$, $\varepsilon < 0$, in the t-plane. As in the preceding paragraph, we can expand $f(t)$ in a Laurent series convergent in this annulus. Setting $t = re^{i\theta}$, we get a real even function $\varphi(\theta) = f(e^{i\theta})$ on the unit circle. Again as in the preceding paragraph, trigonometric interpolation to this at equidistant points $\theta_k = 2\pi k/N$ converges exponentially. But this corresponds to polynomial interpolation to $g(x) = g(\cos \theta) = f(t)$ at the zeros of the Chebyshev polynomials $T_n(x)$, which therefore also converges exponentially (see [7, p. 245], and [V]).

2. Norms. The powerful and flexible mathematical concept of a *norm* on a function space provides a cornucopia of measures of approximation. Each norm defines a distance $\|f - g\|$ in function space, and different norms often correspond to different kinds of convergence. Thus $\|f_n - f\|_\infty \to 0$, $\|f_n - f\|_1 \to 0$, and $\|f_n - f\|_2 \to 0$, are equivalent to uniform, mean, or mean square convergence, respectively.

Given a norm $\|\cdot\|$ and a subclass S of functions, a member f of S such that $\|f - u\| \leq \|s - u\|$ for all $s \in S$ is called the *best approximation* to u in S. The search for best uniform, best mean and best mean square approximations to a given function u is a standard problem in approximation.

For different mathematical and physical problems, very different norms may be appropriate. Thus, for physical problems with a quadratic energy function, such as those of a vibrating string or loaded spring, the square root of the *energy* is often the most satisfactory norm. Mikhlin (Lecture 7, [8, Chap. II]) calls such norms "energy norms." In many stochastic and population problems, $\|p - q\|_1$ may be the appropriate norm. It is very important to choose the "right" norm or norms (or seminorms) in analyzing approximations to physical problems.

For the Ritz and Rayleigh–Ritz methods of Lecture 8, quadratic norms are most convenient. This is because they lead to inner product spaces in which one can find a (the) best approximation by simple *orthogonal projection*. Moreover, the orthogonal projection of a given function f on the subspace spanned by a given basis $\varphi_1, \cdots, \varphi_r$ is easy to compute to within roundoff error; it is especially easy if the φ_i are orthogonal.

Unfortunately, the monomials $1, x, x^2, \cdots, x^n$ form a very ill-conditioned basis for polynomials of degree n. In such cases, it is often desirable to orthogonalize the φ_i. On the interval $[-1, 1]$, this leads from the monomials $1, x, x^2, \cdots, x^n$ to the Legendre polynomials. (Best uniform approximation leads similarly to the Chebyshev polynomials.) For splines, etc., to be discussed below, bases of "patch functions" give not too ill-conditioned matrices.

Sobolev spaces. It will be observed that the energy integrals of Lecture 8 involve derivatives. Some such energy functions like $\int [u^2 + u'^2]\,dx$ give norms which are topologically equivalent to "Sobolev norms." Note that all Sobolev spaces with $p = 2$ are *Hilbert spaces* (complete Euclidean vector spaces), though they are not L_2-spaces. (Here and below, a Euclidean (or inner product) space is a real vector space with an inner product (f, g) having the usual formal properties, including $(f, f) > 0$ for $f \neq 0$.)

Semi-ordered spaces. In other problems, order is relevant, e.g., when dealing with positive linear operators. This is especially true of L_1-norms.[2]

3. Best approximation. It is relatively easy to obtain best mean square approximations in any Euclidean (i.e., inner product) norm by orthogonal projection. Moreover, orthogonal projection A is always *linear*: $A[c_1 f_1 + c_2 f_2] = c_1 A[f_1] + c_2 A[f_2]$.

For example, if

$$f(\theta) = a_0/2 + \sum_{k=1}^{\infty} [a_k \cos k\theta + b_k \sin k\theta],$$

the best mean square approximation to $f(\theta)$ by a linear combination of $1, \cos \theta, \sin \theta, \cdots, \cos n\theta, \sin n\theta$ is simply the usual truncated Fourier series

(3) $$f_n(\theta) = a_0/2 + \sum_{k=1}^{n} [a_k \cos k\theta + b_k \sin k\theta].$$

The f_n always converge to f in the mean square norm $\|f\|_2$ if f is square-integrable (i.e., if $\|f\|_2 < \infty$). For periodic $f \in C^1$, $f_n(\theta) \to f(\theta)$ uniformly in (3). However, for some $f \in C$, we do *not* have $\|f_n - f\|_\infty \to 0$; the f_n of (3) need not converge uniformly, or even pointwise, to f.

---

[2] Direct relations between order and numerical bounds are also derived in L. Collatz, *Numerical Analysis and Functional Analysis*, Springer, Berlin, 1964.

On the other hand, for any $f \in C$ and $\varepsilon > 0$, there always *exists* some trigonometric polynomial of degree $n = n(f, \varepsilon)$:

$$(3') \qquad g_n(\theta) = A_0^n/2 + \sum_{k=1}^{\infty} [A_k^n \cos k\theta + B_k^n \sin k\theta]$$

such that $\|g_n - f\|_\infty < \varepsilon$.

Weierstrass approximation theorem. Likewise, for given $f \in C[-1, 1]$ and $\varepsilon > 0, p(x)$, there always exists an $n = n(\varepsilon)$ and a polynomial $p_n(x)$ of degree n or less such that $\|f(x) - p_n(x)\|_\infty < \varepsilon$. Indeed, by restricting attention to even periodic functions and using the Chebyshev polynomials, one can derive this result from the corresponding fact about trigonometric approximation.[3]

Using the Remez algorithm,[4] one can even compute the polynomial $p_n(x)$ of degree n or less which minimizes $\|f(x) - p_n(x)\|_\infty$ or, equivalently, find the a_k giving

$$\min[\max|f(x) - p_n(x)|] \quad \text{for} \quad p_n(x) = \sum_{k=0}^{n} a_k x^k.$$

One can also use exchange methods or linear programming.[5]

To compute the analogous A_k^n and B_k^n in (3') which will minimize $\|g_n - f\|_\infty = \max|g_n(\theta) - f(\theta)|$ seems to be more awkward.

4. Univariate splines. As already stated in § 1, the error committed in univariate Hermite interpolation of degree $2m - 1$ is $O(h^{2m-j})$ in $f^{(j)}(x)$ (see [4] and [3]).[6] Furthermore, Hermite interpolation is *local*; hence *piecewise polynomial* Hermite interpolation on any subdivided interval is highly accurate.

A useful alternative to Hermite interpolation of odd degree $2m - 1$ is provided by *spline* interpolation of the same degree. This interpolates to given values $y_k = f(x_k)$ at mesh-points, and derivatives $f^{(j)}(a) = \alpha_j$ and $f^{(j)}(b) = \beta_j, j = 1, \cdots, m - 1$, at the endpoints a unique spline function of degree $2m - 1$, in the following sense.

DEFINITION. A spline function of degree $2m - 1$ on a subdivided interval $[a, b]$ is a function $f \in C^{2m-2}[a, b]$ which is a polynomial of degree $2m - 1$ or less in each subdivision.

The error in *cubic* ($m = 2$) and *quintic* ($m = 3$) spline interpolation is $O(h^{2m})$ in the uniform norm. For curve fitting, it has the advantage over Hermite interpolation of not requiring the slopes to be measured (except at the endpoints), but the disadvantage of not being local. Indeed, discontinuities in low order derivatives

---

[3] D. V. Widder, *Advanced Calculus*, Prentice-Hall, Englewood Cliffs, New Jersey, 1947, § 7.2.

[4] See Stiefel in [8], in SIAM J. Numer. Anal., 1 (1964), pp. 164–176, and in [5, pp. 68–82]; also Meinardus in [10, § 7].

[5] J. B. Rosen, SIAM J. Numer. Anal., 7 (1970), pp. 80–103.

[6] For optimal theoretical bounds, see G. Birkhoff and A. S. Priver, J. Math. and Phys., 46 (1967), pp. 440–447; C. A. Hall, J. Approx. Theory, 1 (1968), pp. 209–218; B. Swartz, Bull. Amer. Math. Soc., 74 (1968), pp. 1072–1078.

can adversely affect convergence. As a projector, the (linear) spline interpolation operator has norm $O(h_{\max}/h_{\min})$ in the uniform norm; see [16].

The preceding results are for spline *interpolation*. For applications to elliptic problems, spline *approximation* is more relevant (see Lecture 8). In this connection it is noteworthy that univariate spline interpolation of degree $2m - 1$ is an *orthogonal* approximation method with respect to the inner product $\int f^{(m)}(x) g^{(m)}(x) \, dx$ [1, p. 77], and hence "best" with respect to the norm whose square is $\int |f^{(m)}(x)|^2 \, dx$.

However, there are other good schemes of spline approximation. Very attractive is a linear scheme of "spline approximation by moments," which can be extended (using tensor products) to n dimensions. For any sufficiently smooth function, this method gives a spline approximation of degree $2m - 1$, the error in whose jth derivative is $O(|\pi|^{2m-j})$ for $j = 0, 1, \cdots, 2m - 2$. De Boor and Fix [18] have recently generalized it to a more general family of "spline approximations by quasi-interpolants" (see also [17]).

The Rayleigh–Ritz method, to be discussed in Lecture 8, is still another approximation method; it is especially well-suited to a wide range of physical problems. This can be applied with approximating subspaces of either Hermite or spline functions. As I have stated, spline approximation has the same order of accuracy as Hermite approximation of degree $2m - 1$, with about $1/m$ times as many unknowns. However, the use of splines requires much more programming effort.

For computational efficiency, especially with iterative methods, one wants to have sparse matrices (e.g., of inner products; see Lecture 8). Moreover, for accuracy and numerical stability, one wants the basis of spline functions to be well-conditioned. A good way to achieve both is to use a basis of *patch functions*, whose support consists of relatively few mesh-intervals, e.g., 2 for cubic Hermite and 4 for cubic spline functions. In the case of a uniform mesh, these functions were first determined by Schoenberg.[7] Relative to these bases, ordinary inner products define "band matrices" of widths 7 and 9, respectively.

5. Tensor products. Using tensor products, it is easy to extend the algorithms for Lagrange, Hermite and spline interpolation defined above from univariate interpolation in intervals to bivariate interpolation in rectangles. In particular, tensor products of "patch functions" of one variable, whose support is confined to k successive intervals, have their support confined to a $k \times k$ block of sub-rectangles. However, they need not yield band matrices of fixed finite width.

This tensor product approach gives rise to algebraic existence and uniqueness theorems, of which I shall give only one sample.

THEOREM (de Boor[8]). *Given a rectangle $\mathscr{R}$ and a rectangular subdivision of $\mathscr{R}$, there exists one and only one bicubic spline function which assumes given values at*

---

[7] I. J. Schoenberg, Quart. Appl. Math., 4 (1946), pp. 45–99 and pp. 112–141.
[8] Carl de Boor, J. Math. and Phys., 41 (1962), pp. 212–218; [5, p. 173].

all mesh-points, has given normal derivatives (u_x *or* u_y) *at all edge mesh-points, and given cross-derivatives* u_{xy} *at corner mesh-points.*

What is crucial, moreover, Hermite and spline *interpolation* give good *approximations*: $O(h^4)$ for values and $O(h^{4-j})$ for partial derivatives of order j with bicubic splines, if $u \in C^4(\mathscr{R})$, uniformly and regardless of the number of mesh-points. This contrasts notably with Lagrange (polynomial) interpolation.[9]

In a single rectangle, the error satisfies:

$$\|e^{(r,s)}\| \leq M(\|u^{(4,0)}\| + \|u^{(0,4)}\|)h^{4-\max(r,s)},$$

provided $r, s \leq 4$; the exponents are best possible. The proof is difficult, and due to J. H. Bramble and S. R. Hilbert, Numer. Math. (to appear).

Spline approximating subspaces achieve the same order of accuracy as Hermite approximating subspaces of the same degree (e.g., cubic), with bases about one quarter as big. However, when one uses bases of "patch functions" with minimum support (to get sparse matrices), one finds that the proportion of nonzero entries in the resulting matrices is about four times as great; the net computational advantage of using spline as contrasted with Hermite approximating subspaces seems to be small for solving elliptic DE's.

Bicubic spline interpolants in rectangles can be characterized by a very beautiful variational property [1, p. 242]. Namely, they minimize

(4) $$J[u] = \iint_{\mathscr{R}} [(\partial^4 u)/(\partial x^2 \, \partial y^2)]^2 \, dx \, dy$$

in the class of all smooth functions interpolating to the same conditions (cf. [1, p. 175, (8.6)]). Hence, relative to the quadratic seminorm $\|u\| = (J[u])^{1/2}$, bicubic spline interpolation gives a "best approximation" (see §4).

Unfortunately, as Joyce Mansfield has pointed out, they do not *uniquely* minimize J: if one adds to u any function $v(x)$ which vanishes together with $v'(x)$ and $v''(x)$ at all mesh-points, then $J[u + v] = J[u]$ also minimizes J and solves the same interpolation problem, without in general being a bicubic spline.

6. Rectangular polygons. Bicubic (and biquintic) Hermite interpolation has another advantage: it is *local*. Hence it can be used in (subdivided) rectangular polygons without loss of accuracy (see [2]).

Bivariate spline interpolation (and approximation) in rectangular polygons are much more complicated, and not to be recommended to the inexperienced. I shall indicate here only a few striking results; for details and generalizations, I refer you to the references cited below and to my article in [13].

Let $(\mathscr{R}, \pi)$ be a subdivided *rectangular* polygon, that is, a (rectangular) polygon with sides parallel to the coordinate axes, cut up into subrectangles by straight lines joining pairs of (opposite) edges of $\mathscr{R}$. Define a bicubic spline function

---

[9] For the above and other results, see [13, pp. 185–221]; also [2], and D. D. Stancu, SIAM J. Numer. Anal., 1 (1964), pp. 137–163, and W. Simonsen, Skand. Aktuarietidskr., 42 (1959), pp. 73–89.

$f \in \mathrm{Sp}(\mathcal{R}, \pi, 2)$ as a function $f(x, y)$ which is: (i) a bicubic polynomial in each mesh rectangle, and (ii) a function of class $C^2(\mathcal{R})$.

It follows from (i) and (ii) that $f \in C^{2,2}(\mathcal{R})$.[10] Also, for a rectangular polygon with p reentrant corners and hence $4 + 2p$ sides, the dimension of the bicubic spline subspace $\mathrm{Sp}(\mathcal{R}, \pi, 2)$ is $M + (E - p) + 4$, where M is the total number of mesh-points and E the number of mesh-points on the edges of $\mathcal{R}$. Moreover, Carlson and Hall (op. cit.) have constructed an analytically well-set interpolation scheme for interpolating to appropriate values.

We also have the following analogue of (4) [1, p. 255].

THEOREM (Ahlberg–Nilson–Walsh). *Consider the class* Γ *of all functions* $u \in C^{(2,2)}(\mathcal{R})$ *which satisfy the following conditions in a subdivided rectangular polygon* $[\mathcal{R}, \pi]$:
  (i) $u(x_i, y_j) = f_{ij}$ at all mesh-points (x_i, y_j),
  (ii) $\partial u/\partial y(x_i, y_j) = g_{ij}$ for all (x_i, y_j) on horizontal edges,
  (ii') $\partial u/\partial y(x_i, y_j) = h_{ij}$ for all (x_i, y_j) on vertical edges,
  (iii) $\partial^2 u/\partial x\, \partial y = c_{ij}$ on corners.
Then $u \in \mathrm{Sp}[\mathcal{R}, \pi]$ *implies that* u *minimizes* (4) *in the set* Γ.

This is however only a very *partial* generalization of the corresponding theorem for (subdivided) rectangles, because the greatest lower bound of J need not be attained. To avoid this difficulty, it is simplest theoretically to use the following result.

WHITNEY'S EXTENSION THEOREM. *Any "smooth" function can be extended to all space without losing differentiability.*

As Martin Schultz[11] and Richard Varga [V'] have pointed out, this result makes it possible to justify a priori estimates of the order of accuracy, by reducing to the case of a rectangle. However, no simple *construction* for actually computing such an extension is known to me.

Using this and other observations, Schultz has also shown that, for a significant class of elliptic problems including the Poisson equation with natural (Neumann) boundary conditions, bicubic spline functions on "quasi-uniform" meshes with uniformly bounded mesh-ratios give accuracy of the maximum possible expected order, $O(h^4)$.

Approximation. I should also mention the fact (Runge–Bergman–Walsh) that one can approximate any harmonic function arbitrarily closely by harmonic polynomials, without using *piecewise polynomial* functions at all.

Singular functions. To obtain more accurate approximations to functions which have singularities on the boundaries (e.g., at corners), one must use a subspace of approximating functions which matches those singularities asymptotically. This method has been successfully used by George Fix.[12]

To approximate accurately solutions of elliptic DE's whose coefficients are discontinuous across interfaces with *corners* is even more difficult. The first step

---
[10] Ralph Carlson and C. A. Hall, WAPD-T-2160.
[11] SIAM J. Numer. Anal., 6 (1969), pp. 161–183 and 184–209; both L^∞- and L^2-approximation are discussed.
[12] J. Math. Mech., 18 (1969), pp. 645–658.

consists in determining the asymptotic nature of the singularities which occur at such corners. After much work, some progress has been made on this problem.[13]

REFERENCES FOR LECTURE 7

[1] J. H. AHLBERG, E. N. NILSON AND J. L. WALSH, *The Theory of Splines and their Applications*, Academic Press, New York, 1967.
[2] G. BIRKHOFF, M. H. SCHULTZ AND R. S. VARGA, *Piecewise Hermite interpolation . . . with applications to partial differential equations*, Numer. Math., 11 (1968), pp. 252–256.
[3] P. G. CIARLET, M. H. SCHULTZ AND R. S. VARGA, *Numerical Methods . . . , I. One-dimensional Problems*, Ibid., 9 (1967), pp. 394–430.
[4] PHILIP J. DAVIS, *Interpolation and Approximation*, Blaisdell, Waltham, Massachusetts, 1963.
[5] H. L. GARABEDIAN, editor, *Approximation of Functions*, Elsevier, Amsterdam, 1965.
[6] DUNHAM JACKSON, *The Theory of Approximation*, American Mathematical Society, Providence, 1930.
[7] CORNELIUS LANCZOS, *Applied Analysis*, Prentice-Hall, Englewood Cliffs, New Jersey, 1956.
[8] R. E. LANGER, editor, *On Numerical Approximation*, University of Wisconsin Press, Madison, 1959.
[9] G. G. LORENTZ, *Approximation of Functions*, Holt, Rinehart and Winston, New York, 1966.
[10] G. MEINARDUS, *Approximation of Functions: Theory and Numerical Methods*, Springer, Berlin, 1967.
[11] J. R. RICE, *The Approximation of Functions*, vols. I, II, Addison-Wesley, Reading, Massachusetts, 1964, 1968.
[12] ARTHUR SARD, *Linear Approximation*, American Mathematical Society, Providence, 1963.
[13] I. SCHOENBERG, editor, *Approximation with Special Emphasis on Spline Functions*, University of Wisconsin Press, Madison, 1969.
[14] A. F. TIMAN, *Theory of Approximation of Functions of a Real Variable*, Pergamon-Macmillan, London, 1963.
[15] J. L. WALSH, *Interpolation and Approximation by Rational Functions in the Complex Domain*, 2nd ed., American Mathematical Society, Providence, 1956.
[16] CARL DE BOOR, *On uniform approximation by splines*, J. Approximation Theory, 1 (1968), pp. 219–262.
[17] G. J. FIX AND G. STRANG, *Fourier analysis of the finite element method in Ritz–Galerkin theory*, Studies in Applied Math., 48 (1969), pp. 265–273.
[18] C. DE BOOR AND G. J. FIX, *Spline approximation by quasi-interpolants*, unpublished manuscript.

---

[13] G. Birkhoff, *Angular singularities of elliptic problems*, to appear in J. Approximation Theory, with references to related work of R. B. Kellogg and others.

LECTURE 8

Variational Methods

1. Introduction. In Lecture 2, § 5, I recalled some variational principles which could be used to characterize solutions of boundary value problems of mathematical physics for a number of specific self-adjoint elliptic partial DE's. In particular, every configuration of static equilibrium in classical (Lagrangian) mechanics minimizes a suitable energy function J. In continuum mechanics, J is typically an energy integral whose integrand is quadratic in a "displacement function" $u(\mathbf{x})$ and its derivatives of order up to k. The associated Euler–Lagrange DE equivalent to $\delta J = 0$ is then an elliptic DE of order $2k$.

Such variational principles apply, for example, to the source problem

(1) $$\nabla \cdot [p(\mathbf{x})\nabla u] - q(\mathbf{x})u = f(\mathbf{x}),$$

whose solution minimizes the integral J in

(1') $$\delta \int [p\nabla u \cdot \nabla u + q(\mathbf{x})u^2 + 2f(\mathbf{x})u(\mathbf{x})]|d\mathbf{x}| = 0$$

for given boundary values. Conversely, (1) is the Euler–Lagrange DE associated with the variational problem (1'). In the special case $p = 1$ and $q = f = 0$, this reduces to the Dirichlet principle.

A more sophisticated variational principle refers to a simply supported plate with Poisson ratio v and load density $\rho(x, y)$, resting on the plane $z = 0$. Here the equilibrium condition is $\delta J = 0$, for

(2) $$J[u] = \iint_R \{\tfrac{1}{2}[(\nabla^2 u)^2 - 2(1 - v)(u_{xx}u_{yy} - u_{xy}{}^2)] - \rho u\}\, dx\, dy.$$

As was stated in Lecture 2, § 5, the Euler–Lagrange DE for the minimization of (2) is $\nabla^4 u = 0$, regardless of v: the integral over R of $(u_{xx}u_{yy} - u_{xy}^2)$ is absorbed into a "boundary term" (boundary stress); see also § 8 below.

Again, the surfaces of constant mean curvature spanning a given curve γ (Lecture 1, § 8) are those of least area subject to the constraint of bounding a given total volume (e.g., of a liquid drop).

Similarly, the eigenfunctions of various self-adjoint elliptic DE's are the stationary points of suitable Rayleigh quotients $R[u]$, and are the functions such that $\delta R = 0$ (see Lecture 2 and § 9 below).

The beauty and simplicity of such variational principles has inspired many scientists. They were appealed to by Dirichlet, Riemann and Hilbert for proving existence theorems. And they were utilized by Rayleigh, Ritz, and other scientists to compute *approximate numerical solutions* of elliptic problems which defied exact analysis. I shall describe below some general methods used for this purpose, emphasizing various[1] computational problems which they involve.

These methods are *direct variational methods*, where we give to the word "direct" the special computational significance attributed to it by Sobolev, as quoted by Mikhlin in the preface to [8]: "Direct methods are those methods... which reduce ... problems to the solution of a finite number of algebraic equations."

2. Ritz method. The Rayleigh–Ritz method for finding approximate solutions of variational problems can be described very simply. If $J[u]$ is the functional to be minimized, then one constructs an *approximating subspace* of functions $u(\mathbf{x}, \boldsymbol{\alpha})$, $\boldsymbol{\alpha} = (\alpha_1, \cdots, \alpha_n)$, which depend on n parameters α_j. One then computes

$$(3) \qquad J[u(\mathbf{x}, \boldsymbol{\alpha})] = F(\boldsymbol{\alpha}) = F(\alpha_1, \cdots, \alpha_n)$$

and tries to minimize F in $\boldsymbol{\alpha}$-space.

Given any positive definite *quadratic functional* like those in the examples of the last section, and a linear approximating subspace with basis $f_1(\mathbf{x}), \cdots, f_n(\mathbf{x})$, we obviously have

$$(3') \qquad J[\alpha_1 f_1(\mathbf{x}) + \cdots + \alpha_n f_n(\mathbf{x})] = \tfrac{1}{2} \sum c_{ij} \alpha_i \alpha_j + \sum b_k \alpha_k.$$

To minimize this algebraic expression is straightforward; the condition is

$$(4) \qquad \sum c_{ij} \alpha_j = b_i, \qquad\qquad j = 1, \cdots, N;$$

the solution is unique if the matrix $C = \|c_{ij}\|$ is positive definite (as it is in typical physical problems).

Example 1. Let B be any simply connected domain in the complex z-plane which contains $z = 0$ as an interior point. Then, among all complex analytic functions with $f(0) = 0$ and $f'(0) = 1$, the one which maps B onto a circular *disc* is characterized by either of the following two variational properties:

$$(5) \qquad \delta \iint_B |f'(z)|^2 \, dx \, dy = 0$$

and

$$(6) \qquad \delta \int_{\partial B} |f'(z)|^2 \, ds = 0.$$

Here one can take for the approximating subspace the set of all polynomials $f(z) = z + \sum_{k=2}^{n} c_k z^k$; the Ritz method would approximate the quadratic functional (5) or (6) on that subspace.

In the half-century following Ritz's original paper, many approximating subspaces were used: polynomials (Ritz, [KK]), functions piecewise bilinear in rectangles [3], and functions piecewise linear in triangles [11]. Actually, to minimize

---
[1] See the end of § 3.

the Dirichlet integral in either of the two preceding approximating subspaces is equivalent to postulating that the values at mesh-points satisfy a suitable linear difference equation. Hence, with these approximating subspaces, the Ritz method is equivalent (at mesh-points) to a suitable difference method.

The 5-point ΔE also minimizes a quadratic "energy" functional in the network analogy. Hence the SOR and ADI methods can be regarded as methods for minimizing quadratic functionals on *special* many-dimensional approximating subspaces.

However, the great breakthrough of the past 5–10 years has consisted in the exploitation of *piecewise polynomial* approximating functions, of the kind which I discussed in Lecture 7. These can be combined with quite *general* variational methods, to whose numerical implementation this lecture will be largely devoted.

Because the variational expression (1') involves not only u but also its first derivatives, which are approximated to order $2m - 1$ by piecewise polynomial functions of order $2m - 1$, it is natural to guess that the Rayleigh–Ritz approximation by Hermite or spline functions of these degrees would have an error of the same order. However, it is a remarkable fact that their order of accuracy is actually $O(h^{2m})$ (see Varga's Regional Conference Lectures [V']).

3. Computing the minimum. When the Ritz method is applied to minimize a positive definite *quadratic* functional, the approximate solution minimizes a positive definite quadratic function $F(\alpha)$ of n parameters, which I shall rewrite as $F(\mathbf{x})$ by a simple change of notation which replaces (3) by

$$(7) \qquad F = F_0 + \sum F_i x_i + \tfrac{1}{2} \sum a_{ij} x_i x_j.$$

Here $F_i = \partial F/\partial x_i(\mathbf{0})$, $a_{ij} = \partial^2 F/\partial x_i \, \partial x_j(\mathbf{0})$, and $A = \|a_{ij}\|$ is a positive definite symmetric matrix.

The minimum F occurs where the gradient of F vanishes, or

$$(8) \qquad \sum_{j=1}^{n} a_{ij} x_i + F_i = 0, \qquad i = 1, \cdots, n.$$

This is a linear problem of the form $A\mathbf{x} = \mathbf{b}$ discussed in Lectures 4 and 5, with $b_i = -F_i = -\partial F/\partial x_i(\mathbf{0})$.

It is naturally easier to solve $A\mathbf{x} = \mathbf{b}$ if one uses coordinates with respect to which the Hessian $A = \|\partial^2 F/\partial x_i \, \partial x_j(\mathbf{0})\|$ is a sparse or at least well-conditioned matrix. Such coordinates always *exist*; in fact there always exists a coordinate system making $\partial^2 F/\partial y_i \, \partial y_j = I$, the identity matrix. But to *compute* such a basis is much harder than to solve (8), for large n.

Fortunately, one gets such bases automatically not only from difference approximations (where the "sparsity" is great when the stencil is small), but also from variational methods if one uses "patch bases" of piecewise polynomial functions of the kind described at the end of Lecture 7, § 4.

In such cases, with $n < 1000$ unknowns, it is typically most efficient to solve the system (8) in double precision by one of the *elimination* methods discussed in

Lecture 3, such as Gaussian elimination (with pivoting) or the method of conjugate gradients.[2] In the future, with larger problems having $n > 1000$ unknowns, it seems likely that a combination of elimination with iterative strategies, of the kind discussed at the end of Lecture 4, will be most effective.

4. Nonlinear problems; Newton's method. The preceding method can be used more generally, whenever the gradient ∇F of the function to be minimized is easily computable. Writing

$$(9) \qquad F_i(\mathbf{x}) = \partial F/\partial x_i(\mathbf{x}) = 0, \qquad i = 1, \cdots, n,$$

one obtains a system of equations which are satisfied at all minima of F and, more generally, at all critical points. They are *linear* if and only if the functional F is quadratic.

To solve small systems of mildly nonlinear equations like (9), most authors recommend *Newton's method*.[3] This iterative method is based on a local tangent linearization; thus, as applied to (9), it amounts to solving the simultaneous linear equations

$$(9') \qquad \sum_{j=1}^{n} \frac{\partial^2 F}{\partial x_i \partial x_j}(\mathbf{x}) \Delta x_j = -\frac{\partial F}{\partial x_i}(\mathbf{x}),$$

and then setting $\mathbf{x}^{(r+1)} = \mathbf{x}^{(r)} + \Delta \mathbf{x}$. For F quadratic, since grad F is linear, the first iteration already gives the exact solution (up to roundoff).

In general, Newton's method is well known to essentially square the error at each rth iteration or, rather, to make the error $E(r)$ satisfy $|E(r+1)| \leq M[E(r)]^2$ for some finite M. Hence Newton's method converges extremely rapidly *if* one has a good first approximation to begin with.

However, at each iteration one must solve a system like (7) by elimination, say, so that the method is not very economical in general. Moreover, one must often use extensive trial and error experiments to get near enough to the solution so that Newton's method will converge, e.g., so that $E(r) < 1/M$.

5. Galerkin method. It is well-known that the Ritz method of § 2 is just a special case of the more general Galerkin method, in somewhat the same sense that the theory of orthogonal expansion is just a special case of the theory of biorthogonal expansion. This is explained very well in Kantorovich and Krylov [KK], in Crandall [5] and elsewhere, where it is observed that the Ritz method is only applicable to *self-adjoint* elliptic problems.

Specifically, if $L[u] = f$ is equivalent to $\delta J = 0$ from (3), then one can still apply (9)–(10). Therefore, one can apply Gaussian and other elimination processes to obtain approximate solutions of the given DE, essentially by looking for critical

---

[2] E. Stiefel, Z. Angew Math. Phys., 3 (1952), pp. 1–33; J. W. Daniel, SIAM J. Numer. Anal., 4 (1967), pp. 10–26.

[3] [6]; A. M. Ostrowski, *Solutions of Equations and Systems of Equations*, 2nd ed., Academic Press, New York, 1966; C. G. Broyden, Math. Comp., 19 (1965), pp. 577–593; J. M. Ortega and W. C. Rheinboldt, SIAM J. Numer. Anal., 4 (1967), pp. 171–190; J. Math. Anal. Appl., 32 (1970).

points of the functional J on finite-dimensional submanifolds (which need not be linear subspaces). All that is needed is that the second variation $\delta^2 J$ have eigenvalues bounded away from zero!

However, I have had no personal experience with this method, and so shall say no more about it.

Error bounds. In principle, it is straightforward to give error bounds for the Ritz and Galerkin methods. The general idea is first to select an N-dimensional *approximating subspace* S which approximates every *smooth function* (e.g., the unknown solution u) by some $v \in S$ such that the error $e = v - u$ is "small" in the sense that $J[v] - J[u]$ is *predictably small*. I have given examples of this in Lecture 7; differentiable piecewise bicubic polynomial functions are often effective in practice. Then one must obtain bounds on $\|v - u\|$ (in suitable norms) in terms of $J[v] - J[u]$, say, by bounding the norm of $[\delta^2 J]^{-1}$.

Convexity and monotonicity. A function $F(x_1, \cdots, x_n)$ of finitely many variables is strictly convex in a convex region whenever its Hessian $\|\partial^2 F/\partial x_i \partial x_j\|$ is positive definite there. In a *compact* region, this is equivalent to the following condition on the gradient $\mathbf{G}(\mathbf{x}) = \nabla F(\mathbf{x})$:

$$[G(\mathbf{y}) - G(\mathbf{x})] \cdot |\mathbf{y} - \mathbf{x}| \geq K \cdot |\mathbf{y} - \mathbf{x}|^2, \qquad K > 0.$$

An operator γ on an inner product space which has this property, namely,

(10) $$[\gamma(\mathbf{y}) - \gamma(\mathbf{x})] \cdot |\mathbf{y} - \mathbf{x}| \geq K \cdot |\mathbf{y} - \mathbf{x}|^2, \qquad K > 0,$$

is called *strictly monotone*. In the past 10 years, there has developed a significant existence and uniqueness theory for such strictly monotone operators, which is applicable in particular to the condition $\mathbf{G}(\mathbf{x}) = \mathbf{0}$ and hence to finding the critical points of convex functionals. For this theory, I refer you to [V'].

6. Gradient method. In most variational problems, (9) is not very helpful for locating the minimum of F. It is more efficient to use direct variational methods for minimizing F, thus solving (9) incidentally. Thus, all the "relaxation methods" for solving (9) discussed in Lecture 4 proceed by minimizing some functional.

The best known method for minimizing F is based on the classical "steepest descent" or *gradient method* of Cauchy (1847).[4] This method has many variants, and can be described as attempting to integrate the system $dx_i/dt = -\partial F/\partial x_i$ of ordinary DE's (see Morrey in [1]).

In general, this system may lead only to a *local* minimum, and to locate the absolute minima of nonconvex functions may require random search techniques (based on "random number generators").[5]

---

[4] See G. Forsythe, Numer. Math., 11 (1968), pp. 57–76; Ostrowski, op. cit. supra; J. W. Daniel, Numer. Math., 10 (1967), pp. 123–131. Among earlier papers, note also J. W. Fischbach, Proc. Symposia Applied Math., vol. IV, 1953, pp. 59–72, and P. W. Rosenbloom, Ibid., pp. 127–76.

[5] See H. A. Spang, III, SIAM Rev., 4 (1962), pp. 343–365 (contains an extensive bibliography). Also, A. J. Gleason, Annals Harvard Comp. Lab., 31 (1961), pp. 198–202; J. Kiefer, J. Soc. Indust. Appl. Math., 5 (1957), pp. 105–136.

Because the gradient method requires calculating many partial derivatives, the most rapid way to locate the minimum of a convex function F may be simply to "improve" or "relax" one coordinate (value) at a time.[6] However, such Gauss–Seidel methods converge very slowly for linear and nonlinear problems, as was explained in Lecture 4.

For mildly nonlinear problems, Gauss–Seidel iteration has been successfully used by Ciarlet, Schultz and Varga,[7] who believe it to be more economical than Newton's method. The fact that it always converges more rapidly than the point-Jacobi method has been shown by Rheinboldt and Moré, and by Porsching.[8]

7. Davidon's method. Far superior to the gradient method is the following iterative method due to W. Davidon.[9]

Take a good approximation to the solution for x_0, let $\nabla u(\mathbf{x}_0) = \mathbf{g}_0$, and let $H_0 = I$. Then compute $\mathbf{S}_0 = -H_0 \mathbf{g}_0$; this will be the direction of steepest descent. Now proceed iteratively as follows for specified H_i, x_i and $\mathbf{g}_i = \nabla u(\mathbf{x}_i)$:

(i) Form $\mathbf{s}_i = -H_i \mathbf{g}_i$, and choose $\alpha_i > 0$ so that $u(\mathbf{x}_i + \lambda \mathbf{s}_i)$ is minimized at $\lambda = \alpha_i$; let $\boldsymbol{\sigma}_i = \alpha_i \mathbf{s}_i$.

(ii) Set $\mathbf{x}_{i+1} = \mathbf{x}_i + \boldsymbol{\sigma}_i$, and compute $u_{i+1} = u(\mathbf{x}_{i+1})$ and $\mathbf{g}_{i+1} = \nabla u(\mathbf{x}_{i+1})$, which will be orthogonal to $\mathbf{s}_i$.

(iii) Set $\mathbf{y}_i = \mathbf{g}_{i+1} - \mathbf{g}_i$.

(iv) Set $A_i = \boldsymbol{\sigma}_i \boldsymbol{\sigma}_i^T$, $B_i = -(H\boldsymbol{\sigma}_i)(H\boldsymbol{\sigma}_i)^T$.

(v) Set $H_{i+1} = H_i + A_i + B_i$.

Like the conjugate gradient method, Davidon's method gives the exact minimum of a quadratic function of n variables in n steps, if there is no roundoff. Unlike the conjugate gradient method (applied to ∇u), it will do this even after entering a quadratic region from a nonquadratic region.

The "stopping problem" of knowing when to stop must of course be taken seriously. Moreover, not much numerical experience has been reported on applying Davidon's method to functions of $n > 100$ variables; hence I shall not discuss it further here.

8. Domain constants. In physical and engineering problems, one is often interested less in the detailed behavior of the solution as a function than in some numerical constant associated with it, such as the electrostatic capacity, torsional

---

[6] For linear problems (quadratic functions), such methods can be traced back to Gauss (see [V, p. 24]). For nonlinear problems, it was proposed by G. Birkhoff and J. B. Diaz, Quart. Appl. Math., 13 (1956), pp. 431–453.

[7] Numer. Math., 9 (1967), pp. 394–430; see also J. Ortega and M. Rockoff, SIAM J. Numer. Anal., 3 (1966), pp. 497–513.

[8] See J. Ortega and W. Reinboldt, *Iterative Solution of Nonlinear Equations in Several Variables*, Academic Press, New York, 1970, and the references given there; T. A. Porsching, SIAM J. Numer. Anal., 6 (1969), pp. 437–449.

[9] R. Fletcher and M. J. D. Powell, Comput. J., 6 (1963), pp. 163–168, give a very readable exposition; W. Davidon, Ibid., 10 (1968), pp. 406–410, discusses it. See also C. G. Broyden, Math. Comp., 21 (1967), pp. 368–381; J. Greenstadt, Ibid., pp. 360–367; A. A. Goldstein and J. F. Price, Numer. Math., 10 (1967), pp. 184–189; Y. Bard, Math. Comp., 22 (1968), pp. 665–666; [13, pp. 1–20] and [15, pp. 43–61].

rigidity, hydrodynamical virtual mass, fraction of wave energy scattered, or lowest eigenvalue. These constants may be called *domain constants*; many of them can also be described by variational principles; thus the electrostatic capacity is the minimum self-energy

$$\int\int de(\mathbf{x})\, de(\xi)/|\mathbf{x} - \xi| \text{ for } \int de(\mathbf{x}) = 1$$

(Gauss principle). See [10], [K], etc.

It is natural to guess that such domain constants vary continuously, or even monotonically, with the domain. This is usually so, and is a very helpful fact. However, the following is a remarkable exception.

Babuška paradox. Let Δ be the unit disc, and let Δ_n be the regular n-sided polygon inscribed in Δ. Consider the following boundary value problem:

(11) $\qquad \nabla^4 u = 1 \text{ in } R, \quad u = \partial^2 u/\partial n^2 = 0 \text{ on } \partial R.$

If $\varphi(\mathbf{x})$ is the solution of (11) for $R = \Delta$, and $\varphi_n(\mathbf{x})$ its solution for $R = \Delta_n$, one would expect to have the $\varphi_n(\mathbf{x})$ approach $\varphi(\mathbf{x})$ as $n \to \infty$. The Babuška paradox consists in the fact that they do not.

A partial explanation of the Babuška paradox is provided by the comment after (2) (see Lecture 7, [13, § 9] for a further explanation).[10]

9. Rayleigh quotient. The eigenfunctions of many classical vibration problems are those functions which define *critical points* of the Rayleigh quotient. This is defined as the ratio

$$R[u] = N[u]/D[u] = \text{(potential energy)/(kinetic energy)},[11]$$

(cf. Lecture 2, (11)). The lth eigenfunction φ_l is then generally characterized by the following *minimax property*: $R[\varphi_l]$ minimizes the value of $R[u]$ on the subspace orthogonal to $\varphi_1, \cdots, \varphi_{l-1}$; $R[\varphi_l] = \lambda_l$ is the lth eigenvalue; and the l-dimensional subspace spanned by $\varphi_1, \cdots, \varphi_l$ minimizes the maximum of $R[c_1\varphi_1 + \cdots + c_l\varphi_l]$, considered as a functional on l-dimensional subspaces.

Equivalently, the φ_l are the critical ("stationary") points of the potential ("strain") energy on the unit sphere in the Hilbert space defined by the norm $D[u]^{1/2}$ associated with the kinetic (inertial) energy.

In finite-dimensional approximating subspaces, relative to any basis, this corresponds to finding the critical points of a "discrete Rayleigh quotient," of the form

(12) $\qquad\qquad\qquad R[\boldsymbol{\alpha}] = \boldsymbol{\alpha}B\boldsymbol{\alpha}^T/\boldsymbol{\alpha}A\boldsymbol{\alpha}^T,$

---

[10] See E. Reissner in [15, pp. 79–94] for an analysis of the boundary conditions and their rationale.
[11] Rayleigh, Proc. London Math. Soc., 4 (1873), pp. 357–358; H. Poincaré, Amer. J. Math., 12 (1890), pp. 211–294, § 2; [CH].

where A and B are positive definite and symmetric. This is an algebraic eigenvalue problem, techniques for whose solution are discussed in [W]; it gives *upper bounds* to the true eigenvalues.

A rather complete review of approximate solutions of elliptic eigenvalue problems by variational methods has recently been given by George Fix and the author in (BV, pp. 111–51], together with an extensive bibliography. I shall supplement this review at the end of the next lecture.

Error bounds. One can compute error bounds for the φ as in [2] by first establishing approximation theorems for the denseness of the *approximating* subspace. These are typically of the form

$$\|v - u\| \leq K\|v\|_\gamma N^\beta,$$

where N is the dimension of the approximating subspace, and may refer to the numerator *or* the denominator of the Rayleigh quotient. By combining these, one can obtain bounds on the errors of the $\lambda_l = R[\varphi_l]$; they typically involve *Sobolev norms.* A favorable point is the fact that the error in λ_l is proportional to the *square* of the error in $R[u]$.

Lower bounds. The Rayleigh–Ritz method gives *upper* bounds to eigenvalues. Lower bounds can also be computed by methods due to Weinstein and developed for computational problems by Bazeley (see [7] and [6]).[12]

Adding inertia lowers all eigenvalues of any mechanical system, while adding stiffness increases them. The algebraic counterpart of this principle of mechanics is Weyl's monotonicity theorem: adding a positive definite symmetric matrix to B increases all eigenvalues; adding such a matrix to A decreases them.

REFERENCES FOR LECTURE 8

[1] E. F. BECKENBACH, editor, *Modern Mathematics for the Engineer*, McGraw-Hill, New York, 1956.
[2] G. BIRKHOFF, C. DE BOOR, B. SWARTZ AND B. WENDROFF, *Rayleigh–Ritz approximation by piecewise cubic polynomials*, SIAM J. Numer. Anal., 3 (1966), pp. 188–203.
[3] R. COURANT, *Variational methods for the solution of problems of equilibrium and vibrations*, Bull. Amer. Math. Soc., 49 (1943), pp. 1–23.
[4] ———, *Dirichlet's Principle*, Interscience, New York, 1950.
[5] S. H. CRANDALL, *Engineering Analysis*, McGraw-Hill, New York, 1956.
[6] GEORGE FIX, *Orders of convergence of the Rayleigh–Ritz and Weinstein–Bazley methods*, Proc. Nat. Acad. Sci. U.S.A., 61 (1968), pp. 1219–1223.
[7] S. H. GOULD, *Variational Methods for Eigenvalue Problems*, University of Toronto Press, Toronto, 1966.
[8] S. G. MIKHLIN, *Variational methods in Mathematical Physics*, Macmillan, 1964.
[9] C. B. MORREY, *Multiple Integrals in the Calculus of Variations*, Springer, Berlin, 1966.
[10] G. PÓLYA AND G. SZEGÖ, *Isoperimetric Inequalities in Mathematical Physics*, Princeton University Press, Princeton, 1951.
[11] J. L. SYNGE, *The Hypercircle in Mathematical Physics*, Cambridge University Press, London, 1957.

---

[12] Also J. B. Diaz, Proc. Symposia Applied Math., vol. VIII, American Mathematical Society, 1956, pp. 53–78, and [BV, pp. 145–50]; A. M. Arthurs, *Complementary Variational Principles*, Oxford University Press, 1970; B. L. Moiseiwitsch, *Variational Principles*, Interscience, 1966.

[12] J. W. DANIEL, *Theory and Methods for Approximate Minimization*, Prentice-Hall, Englewood Cliffs, New Jersey, 1970.
[13] R. FLETCHER, editor, *Optimization*, Academic Press, New York, 1969.
[14] M. J. D. POWELL, *A survey of numerical methods for unconstrained optimization*, SIAM Rev., 12 (1970), pp. 79–97.
[15] *Studies in Optimization* 1, SIAM Publications, 1970.
[16] S. G. MICHLIN, *Numerische Realisierung von Variationsmethoden*, Akademie-Verlag, Berlin, 1969. (Contains references to the Soviet literature.)

LECTURE 9

Applications to Boundary Value Problems

1. General remarks. The preceding lectures emphasized theoretical ideas and results relating to the numerical solution of elliptic problems. I shall conclude by summarizing some impressions of the current "state of the art" of solving elliptic problems numerically, as applied by engineers, physical chemists and other *users*. These impressions are based partly on browsing in journals published by the ASME, IEEE, American Institute of Physics, etc., and partly on my own participation in the efficient solution of a handful of substantial specific problems. Here by "substantial" I mean that the final program was the fruit of 5–25 man-years of cooperative effort.

Users have very different viewpoints from numerical analysts, or even specialists in the broader field of scientific computing. In the first place, they rely minimally on mathematical theory. Even simple theorems which bear directly on the problem to be solved are less convincing to them than numerical evidence. Of minimal or negative interest to them are complicated theorems designed to give an appearance of enormous sophistication and generality, but whose hypotheses cannot be easily tested in specific cases.

Secondly, users seldom want massive *tables*; the problem of economical information storage, retrieval, and reading is a major one. High-speed computers can print out more 10-digit numbers in a minute than a human user can evaluate in a day. Though graphical output is easier to scan, even this may contain too much information. Users may want to know only a few integral parameters such as the heat transfer rate, growth rate, electrostatic capacity, lowest eigenvalue, or some quantity such as the peak stress or peak temperature which must be limited for reasons of safety.

Thirdly, users tend to regard computer printouts as revealing only one aspect of a much larger picture. This is because most computations are based on highly simplified mathematical *models*, which usually make drastic approximations and ignore many variables which affect the real situation. Thus, for elliptic problems, the capacity of current computers is severely taxed by even *three* independent variables, whereas most elliptic problems of quantum mechanics involve six or more variables.

Finally, users can usually get the information which they want more efficiently by *combining* numerical methods of the kind described above with *analytical* methods, than by arithmetic ("number-crunching") alone. Thus, when the terms of an elliptic equation vary by orders of magnitude, it is often most efficient to use *asymptotic* methods.

I shall now illustrate the preceding general remarks about scientific and engineering computing by a hasty and very superficial literature survey. Its purpose is to at least indicate the fascinating variety and complexity of this area. It is an area in which man has so far made only a very small dent. There remain unlimited opportunities in scientific and engineering computing for future workers having sufficient imagination and more powerful computers!

2. Potential flows. I shall begin with the potential flows of classical hydrodynamics (Lecture 1, § 2) and their generalizations. These are the flows which received the most attention from "classical" numerical analysts, partly because they are the flows about which classical analysis gives the most information.

Ideal plane flows past impermeable obstacles can be treated by conformal mapping methods (see Lecture 6 and Lecture 8); so can the flows past wing sections (two-dimensional airfoil theory). For the latter, interest is concentrated on "thin wings" (slightly cambered sections with rounded leading edge and sharp trailing edge). The engineering implications of the mathematical model were pretty well understood[1] by 1917, and no longer of practical interest by the end of World War II, before high-speed computers became available.

Somewhat more difficult is the effective calculation of the potential flow past a solid of revolution of given shape (idealized airship or submarine). The difference $u = \varphi - x$ between the velocity potential $\varphi(x, y, z)$ for the flow of interest and that x for uniform flow is a solution of the Neumann problem in the exterior of the solid for the boundary condition $\partial u/\partial n = \cos \gamma$. Von Kármán proposed calculating it by Rankine's method (1871) of superposing on a uniform flow the potential flows of dipoles distributed along the axis. If one takes the density of this distribution as an unknown function, one gets an integral equation which has been solved by iteration.[2]

Especially since 1940, there has also been considerable interest in "cavity flows" bounded by free streamlines at constant pressure and velocity. However, the calculation of such flows is much more difficult.[3] Even in two dimensions, it involves solving a nonlinear integral equation. In the axially symmetric case or when the effects of gravity are included, the problems become extremely difficult. Interesting work on such potential flows with free boundaries has recently been done by P. R. Garabedian, T. Y. Wu and others.

Although the preceding types of flows have great mathematical interest, they ignore an entire range of physical phenomena involving the effects of viscosity on the development of boundary layers, flow separation and turbulence [6]. I shall say something about computations including these effects in § 6.

---

[1] See the German edition of Lamb, Teubner, 1931, edited by von Mises, who republished his 1917 papers there as Anhang IV; also S. Goldstein, J. Aero. Sci., 15 (1948), pp. 189–220.

[2] L. Landweber, Taylor Model Basin Rep. 761 (1951); J. P. Moran, J. Fluid Mech., 17 (1964), pp. 285–304.

[3] See G. Birkhoff, *Jets, Wakes, and Cavities*, Academic Press, New York, 1957, Chap. IX, and references given there. Also Rep. ACR-38, *Second Symposium on Naval Hydrodynamics*, Washington, 1960, pp. 261–276.

3. Related problems. The Laplace equation and its generalizations such as the Poisson and biharmonic equations arise in many other physical problems. Indeed, a substantial fraction of the numerical solutions of elliptic equations which were published before 1950 concerned these; for example, this is true of all the numerical results in the books by Southwell (Lecture 2, [9]) and Synge (Lecture 8, [11]). In the rest of this section, I shall give a few more examples of such problems from (nonviscous) fluid mechanics; for some other applications, see [13].

Progressive waves. Gravity waves which progress without change of form can also be considered as steady potential flows (hence as solutions of the Laplace equation) with a free surface, relative to moving axes; and an enormous amount of ingenuity has been expended in studying them. However, to date, analytical methods have proved more powerful than numerical methods for analyzing them, and relatively few accurate numerical studies have been reported.[4]

Ship wave resistance. A related problem of great interest concerns the gravity wave resistance (e.g., in still water) to the forward motion of a surface ship. For a so-called "thin ship" (i.e., in the linearized or perturbation approximation) in an ideal fluid, this can be calculated as a quintuple integral by numerical quadrature. However, so far, analytical and semi-analytical methods seem to have been more effective than purely numerical methods.[5] Moreover computed results ignore viscosity, and hence boundary layer and wake resistance. Thus there is probably greater naval interest today in *acceleration* potentials.[6]

Subsonic flow. The flow past an obstacle moving at subsonic speed through an inviscid incompressible fluid has a velocity potential which is governed by a nonlinear partial differential equation which tends to the Laplace equation as the Mach number tends to zero, and which seems to define a well-set problem.[7] A number of interesting attempts have been made to solve this equation by numerical methods,[8] but the methods tested so far seem not to give very accurate results for general problems.

4. Polygonal plates. As was mentioned in Lecture 3, § 6, the equations of elastic equilibrium can be discretized in various ways by difference approximations. Such difference approximations seem especially attractive in the case of *polygonal plates*, because the boundary conditions can then often be applied directly to mesh-points. Indeed, quite a few interesting problems were solved

---

[4] See L. M. Milne-Thomson, *Hydrodynamics*, 5th ed., Macmillan, 1968; E. V. Laitone, J. Fluid Mech., 9 (1960), pp. 430–444; C. Lenau, Ibid., 26 (1966), pp. 309–320.

[5] See G. Birkhoff, B. V. Korvin-Kroukovsky and J. Kotik, Trans. Soc. Naval Arch. Marine Eng., (1954), pp. 359–396. Also, T. Havelock, *Collected Papers on Hydrodynamics*, Publication ONR/ACR-103, U.S. Government Printing Office, 1964.

[6] See J. N. Newman, Annual Revs. Fluid Mech., 2 (1970), pp. 67–94.

[7] R. Finn and D. Gilberg, Comm. Pure Appl. Math., 10 (1957), pp. 23–63; R. Finn, Proc. Symposia Pure Math., vol. IV, American Mathematical Society, 1961, pp. 143–148.

[8] See M. Holt, editor, *Basic Developments in Fluid Dynamics*, Academic Press, New York, 1965; G. S. Roslyakov and L. A. Chudov, *Numerical Methods in Gas Dynamics*, Israel, 1966; Krzyblwocki, Chap. XV, which deals with Bergman's method.

numerically with good (1 %) accuracy in this way before high-speed computers were available, both by difference and by variational methods (see [KK, pp. 215, 286, 322, 595], also Mikhlin).

In treating such problems, care should be taken to make the discrete approximations to boundary conditions retain the self-adjoint character of the exact problem. The safest way to do this is probably to derive the discretization from an approximate (discretized) expression for the energy integral, as was recommended by Stiefel et al. and by Griffin and Varga.[9] These authors were also among the first to develop computer codes for solving elasticity problems. Their codes used block relaxation methods described in Lecture 4 to solve the resulting system of equations; however, it seems likely that Gauss–Choleski elimination would have been about as efficient.

Singularities. The most serious criticism which has been levelled at the application of finite difference methods to solve problems in elasticity concerns their failure to simulate stress concentrations near corners, and especially near notches where they can theoretically become infinite.[10] The practical importance of this discrepancy between difference approximations and the more accurate continuum model depends on the aim of the computations and on the way in which the structure is loaded.

5. Finite-element methods. Over the past 15 years, experts in structural mechanics have developed a variety of so-called *finite-element methods* for the numerical discretization of boundary value problems, which can be regarded as direct applications of the variational and approximation techniques which I have reviewed in the preceding two lectures. Their use is spreading like wildfire. Among the recent books dealing with finite-element methods, I recommend particularly the authoritative expositions by Argyris [1] and Zienkiewicz [11]. Briefer surveys directed to mathematicians have been published by Felippa and Clough and by Pian in [BV, pp. 210–271].[11]

The finite-element approach consists in approximating energy integrals as in Lecture 7, by linear combinations of a basis of compatible "patch functions" defined in terms of the displacements of "nodal points" or "joints" by suitable interpolation formulas. The simplest such functions, and those most commonly used, are the piecewise linear and piecewise bilinear functions in triangular and quadrilateral elements, respectively. In linear elasticity theory, the elastic energy of distortion is a quadratic functional defined on the space of (compatible) linear

---

[9] Engeli, Ginsburg, Rutishauser and Stiefel, Mitt. Inst. ang. Math. #8, Birkhauser Verlag, 1959; Griffin and Varga, J. Soc. Indust. Appl. Math., 11 (1963), pp. 1046–1062.

[10] See the discussion of the Babuška paradox in Lecture 8, § 8; also SIAM J. Appl. Math., 14 (1966), pp. 414–417.

[11] See also B. F. de Veubeke, *Upper and lower bounds for matrix structural analysis*, Pergamon Press, 1964; R. J. Melosh, AIAA J., 1 (1963), p. 1631; I. Holland and K. Bell, editors, *Finite Elements in Stress Analysis*, Tapir, 1969.

combinations of displacements by "stiffness matrices." There exists a production code DUZ-1 based on the above ideas.[12]

Evidently, finite-element methods involve the variational ideas discussed in Lecture 8. In particular, one can develop higher order finite-element methods based on piecewise cubic and piecewise bicubic polynomial displacement functions (e.g., bicubic Hermite approximations).[13] One can moreover prove that such higher order methods have a higher order of convergence, and indeed they are recommended by those who have tried them.

However, the applications of finite-element methods to solid mechanics involve much more than variational ideas and general approximation theory. To apply them successfully, one must be familiar with the various differential equations and integral relations which were derived analytically by "classical" applied mathematicians, and one may wish to use "physical" approximations such as the "lumped mass" approximation [BV, p. 239]. Finally, their adaptation to problems of plasticity, for which no general variational principles are available, will surely require even greater analytical ingenuity and mechanical intuition.

6. Incompressible viscous flows. The discussion of §§ 2–5 concerned primarily linear problems of continuum mechanics. Though few of them have been studied in depth by numerical analysts, existing methods should be adequate for handling them. I shall now take up in more detail typical nonlinear elliptic problems from fluid mechanics, whose numerical analysis by rigorous approximation theorems promises to be a much more formidable task.

One such problem, not governed by a classical variational principle, concerns the steady flow of an incompressible fluid having a specified velocity at infinity, around a solid obstacle S. The mathematical problem is to determine a vector field $\mathbf{u}(\mathbf{x})$ (the velocity field) which satisfies the following nonlinear time-independent Navier–Stokes DE's:

(1) $$\nabla \cdot \mathbf{u} = 0, \quad \mathbf{u} \cdot \nabla \mathbf{u} = \nu \nabla p + \mu \nabla^2 \mathbf{u},$$

subject to the boundary conditions

(2) $$\mathbf{u}(\mathbf{x}) = \mathbf{0} \quad \text{on} \quad \partial S, \quad \lim_{x \to \infty} \mathbf{u}(\mathbf{x}) = (v_\infty, 0, 0);$$

here v_∞ is the "free stream velocity," and S is given.

This problem has been the subject of an enormous amount of theoretical and experimental research. Until recently, even the existence and uniqueness of solutions had not been established theoretically, and many a plausible mathematical idea had proved inadequate to explain the complexities of reality.[14]

---

[12] D. S. Griffin, R. B. Kellogg, W. D. Peterson and A. E. Sumner, Jr., Rep. WAPD-TM-555, Bettis Atomic Power Laboratory.

[13] [11, Chap. 7]; F. K. Bogner, R. L. Fox and L. A. Schmit, Proc. Conference Matrix Methods in Structural Mechanics, AFIT, Wright-Patterson AFB, Ohio, 1965; A. L. Deak and T. H. H. Pian, AIAA J., 5 (1967), pp. 187–189; B. H. Hulme, Doctoral Thesis, Harvard University, 1969.

[14] See G. Birkhoff, *Hydrodynamics: A Study in Logic, Fact, and Similitude*, 2nd ed., Princeton University Press, Princeton, 1950, Chap. 2; also [6].

For an authoritative discussion of existence and uniqueness theorems about such flows, I refer you to the recent book by Mme. Ladyzhenskaya;[15] experts (of whom I am not one!) seem to agree that the boundary value problem defined by (1)–(2) is mathematically well-set.[16]

Physically, however, the behavior of such flows (including their stability) depends dramatically on the *Reynolds number* $R = u_\infty d/v$, where d is the diameter of S. For $R > 1000$, although time-independent solutions of the Navier–Stokes equations may exist mathematically, they are unstable physically; they are never observed.

As $R \downarrow 0$, axially symmetric solutions of (1) approach a limiting so-called *creeping flow* whose "stream function" $\psi(x, r)$, $r = \sqrt{x^2 + y^2}$, satisfies the elliptic DE

$$(3) \qquad vE^4\psi = r\frac{\partial(\psi, r^{-2}E^2\psi)}{\partial(x, r)}, \qquad E^2 = \frac{\partial^2}{\partial x^2} + \frac{\partial^2}{\partial r^2} - \frac{1}{r}\frac{\partial}{\partial r}.$$

Here $\partial(u, v)/\partial(x, y)$ is the Jacobian $u_x v_y - v_x u_y$. The velocity (u_1, u_2) in any (meridian) plane through the axis of symmetry can be computed from ψ by the formulas

$$(3') \qquad u_1 = r^{-1}\partial\psi/\partial r, \qquad u_2 = -r^{-1}(\partial\psi/\partial x).$$

Many applied mathematicians have tried their hand at computing time-independent plane and axially symmetric solutions of the Navier–Stokes equations at low and moderate Reynolds numbers.[17] The case of the plane flow past a *circular cylinder* has received particular attention; a number of references are cited in [2, p. 260, ftnt. 6].[18] In the two-dimensional case, one can introduce a stream function $V(x, y)$ such that

$$(4) \qquad \partial V/\partial x = -v, \qquad \partial V/\partial y = u;$$

this reduces the system of two simultaneous DE's (1) to the single nonlinear elliptic equation

$$(5) \qquad v\nabla^4 V = \partial(V, \nabla^2 V)/\partial(x, y);$$

for the detailed derivation of (3) and (5), see Lamb (Lecture 1, [5]). From the standpoint of fluid mechanics, *the results are inconclusive* for reasons which I shall now try to explain.

---

[15] O. A. Ladyzhenskaya, *The Mathematical Theory of Viscous Incompressible Flow*, 2nd ed., Gordon and Breach, New York, 1969.

[16] R. W. Finn, Arch. Rational Mech. Anal., 25 (1967), pp. 1–39 and 19 (1965), pp. 363–406; see also Acta Math., 105 (1961), pp. 197–244; G. Prodi, Ann. Mat. Pura Appl., 48 (1959), pp. 173–182.

[17] See, for example, Carl Pearson, J. Fluid Mech., 21 (1965), pp. 611–622 and 28 (1967), pp. 323–337; (laminar) natural convection has also been computed by S. W. Churchill and J. D. Hellums, A.I. Chem. E.J., (1962), p. 690. A review article (mostly concerned with initial value problems) has been written by H. W. Emmons [6].

[18] See especially D. N. de G. Allen and R. V. Southwell, Quart. J. Mech. Appl. Math., 8 (1955), pp. 129–145; a more recent computation is that of H. B. Keller and H. Takami [7, pp. 115–127]; see also D. Greenspan, R-5000, Mathematics Research Center, U.S. Army, University of Wisconsin, Madison, 1968.

First, we have the Stokes paradox,[19] which states the unpleasant fact that the two-dimensional problem has *no solution* in the limiting case R = 0 (when its DE reduces to $\nabla^4 V = 0$), whereas the axially symmetric problem has an exact analytical solution first derived by Stokes.

Second, physical interest centers around the phenomenon of *flow separation*, and (although the contrary is asserted in [9]) I do not think that the point where this occurs has been determined in a very convincing way by existing computations (see § 6).

Third, when R > 60 (the number is sensitive to many influences described in [2, Chap. XIII]), the steady flow whose existence has been proved by Leray and others must be unstable; the *stable* flow regime involves periodic vortex shedding. In some remarkable calculations of *time-dependent* solutions of the Navier–Stokes equations, Fromm[20] has been able to simulate this phenomenon *qualitatively* in a very convincing way for a cylinder having a *square* cross-section. However, the computations are themselves sensitive to various influences whose operation is not fully understood, and they have not (as yet) exactly reproduced all the experimental facts; hence the problem is still not solved rigorously.

7. Boundary-layer calculations. In 1904, a major breakthrough in the mathematical analysis of incompressible viscous flows was made by L. Prandtl.[21] On the basis of various intuitive assumptions, Prandtl concluded that as R → ∞ the fluid near a flat plate should satisfy asymptotically the following boundary-layer equations:

(6) $$uu_{xx} + vu_y = -\rho^{-1}p_x + vu_{yy}, \qquad u_x + v_y = 0,$$

where $p = p(x)$ is supposed known. The appropriate boundary conditions are (writing $\lim_{y \to \infty} u(x, y) = U(x)$):

(6') $$u(x, 0) = v(x, 0) = u, \qquad UU_x = -\rho^{-1}p_x.$$

These equations can be derived from the Navier–Stokes equations of § 6 as a singular perturbation.[22]

More generally, the boundary-layer equations (6)–(6') are valid near walls when the boundary-layer thickness is much less than the wall curvature. They were integrated numerically[23] with desk machines 40 years ago, more accurately than the full Navier–Stokes equations are integrated today. However, attempts[24] to calculate the separation point are not completely rigorous because they ignore

---

[19] [1, p. 44]; R. W. Finn and W. Noll, Arch. Rational Mech. Anal., 1 (1957), pp. 95–106.

[20] J. E. Fromm and F. H. Harlow, Physics of Fluids, 6 (1963), p. 975.

[21] Proc. III Internat. Math. Congress, Heidelberg 1904. Note that Prandtl made his initial presentation to an audience of mathematicians, not to physicists or engineers!

[22] [4, § 4.2]; [10, Chap. VII]. For earlier studies, see Lamb–von Mises, p. 812, and S. Goldstein, *Modern Developments in Fluid Dynamics*.

[23] S. Goldstein, Proc. Cambridge Philos. Soc., 26 (1930), p. 1.

[24] See [9, pp. 137, 222]; [10, p. 160]; S. Goldstein, Quart. J. Mech. Appl. Math., 1 (1948), pp. 43–69; and pp. 377–436 of Holt, op. cit. in footnote 8.

the physics of real wakes [2, Chaps. XIII–XIV], and especially the phenomenon of turbulence.

Lubrication calculations. Further asymptotic simplifications of the Navier–Stokes equations are made in the hydrodynamical theory of lubrication. Moreover, a vast amount of useful computation has been based on the resulting equations, which Osborne Reynolds guessed when he founded the hydrodynamical theory of lubrication in 1886. If $h(x, z)$ denotes the clearance gap, and $p(x, z)$ the pressure, these equations are (in an incompressible fluid):

$$\nabla \cdot \left(\frac{h^3}{\mu} \nabla p \right) = 6K \frac{\partial h}{\partial x}, \qquad \mu = \text{viscosity}. \tag{7}$$

Because $h(x)$ typically varies by orders of magnitude in a heavily loaded bearing, this elliptic equation should be integrated by special methods not discussed in this book, but familiar to specialists in lubrication theory.[25]

8. Other problem areas. A study in depth of other areas of physics and engineering would no doubt reveal similar complexity and wealth of phenomenological detail; there is certainly no shortage of challenging problems! I shall conclude my survey of the numerical solution of elliptic DE's by giving you fleeting glimpses of three such problem areas arising in electromagnetic theory, physical chemistry, and nuclear reactor theory, respectively.

Microwave transmission. In general, quantitative predictions of the propagation and scattering of electromagnetic waves still rely primarily on analytical ideas and techniques, of the kind I discussed in Lecture 2. An area which has received particular attention because of the practical importance of radar concerns microwave transmission in waveguides and associated scattering phenomena. Here numerical methods have finally begun to make a dent, especially for treating problems in which the Helmholtz equation $\nabla^2 \phi + k^2 \phi = 0$ plays a central role, for example, in determining the TE-modes and TM-modes of cylindrical waveguides of arbitrary cross-section. Those interested in learning more about this nascent field of scientific computing should study [12] and the references given there; it would seem desirable to apply to it systematically the techniques which I have reviewed in Lectures 2–8.

Schrödinger equation. In spite of minor (e.g., relativistic and nuclear) perturbations which are ignored by the Schrödinger equation

$$\nabla^2 \psi + \frac{8\pi^2 \mu}{h^2} [E - V(\mathbf{x})] \psi = 0, \tag{8}$$

it seems hopeful that most problems of physical chemistry could be solved with

---

[25] W. A. Gross, *Gas Film Lubrication*, John Wiley, New York, 1962; O. Pinkus and B. Sternlicht, *Theory of Hydrodynamic Lubrication*, McGraw-Hill, New York, 1961. For a careful discussion of asymptotics, see A. B. Taylor, Proc. Roy. Soc. Ser. A, 305 (1968), pp. 345–361; for the important case of partial lubrication, see [5, pp. 102–121].

sufficient accuracy if we could *solve* its generalization to n nuclei and electrons:

$$\left(\sum_{k=1}^{n}\frac{1}{\mu_k}\nabla_k^2\right)\Psi + \frac{8\pi^2}{h^2}[E - V(\mathbf{x}_1,\cdots,\mathbf{x}_n)]\Psi = 0. \tag{9}$$ [26]

At least, many physicists and physical chemists have proceeded on this assumption, and the relevant literature is vast.

Perhaps because the domain involved is $3n$-dimensional infinite space, and the potential function V becomes singular when $\mathbf{x} = 0$ in (8) (for the two-body problem of the one-electron hydrogen atom), and on the $(3n - 3)$-dimensional locus $\prod_{i<j}|\mathbf{x}_i - \mathbf{x}_j|^2 = 0$ in (9), *classical analysis* and *physical approximations* (e.g., to an "electron cloud") have played a central role in approximate numerical solutions of the Schrödinger equation, so far. However, it would again seem desirable to see what the methods described in Lectures 2–8 might contribute to supplementing the many ingenious techniques already developed by physical chemists for obtaining approximate solutions of (8) and (9). I would advise anyone who wishes to take up this challenge to study the review articles [3] and [8].

Multigroup diffusion equations. The multigroup diffusion equations of nuclear reactor theory have a different status. Far from having been ignored, many of the methods which I have described were actually *developed* in this context; this is especially true of the iterative and semi-iterative methods for solving very large systems of linear equations which I have reviewed in Lecture 4 and the beginning of Lecture 5.[27] (The same is true of petroleum reservoir exploitation, for which the ADI methods reviewed in Lecture 5 were developed.)

Here a fascinating question concerns the relative merits of the difference methods of Lectures 3–5 and the "finite element" methods which I briefly sketched in Lectures 7–8. Since I am actively working on this question currently, and hope to report my findings elsewhere very soon, I shall say no more about it here.

9. Future perspectives. It is now 25 years since the first automatic large-scale computer became operative, and 20 years since von Neumann made his notable contributions to scientific computing. In this time, computers have become many times more powerful and many large codes for solving scientific problems have been developed. It is natural to ask what the future holds.

In trying to answer this question, it seems relevant to review the progress of the past 20–25 years in those fields which have received the most attention. Among these are included, not unnaturally, the fields in which von Neumann himself acted as a contributor and catalyst. The fields were:[28] three-dimensional turbulence, three-dimensional compression waves with shocks, weather forecasting, chemical and reactor kinetics of moving materials, and the Schrödinger equation.

---

[26] E. C. Kemble, *The Fundamental Principles of Quantum Mechanics*, McGraw-Hill, New York, 1937, Chap. I, §§ 5–7.

[27] Much more complete expositions are given in [V] and [W]; see also [14].

[28] J. von Neumann, *Collected Works*, Pergamon Press, London, 1963, especially vol. V, p. 236, pp. 241–243, and vol. VI, pp. 413–430.

Of these, only the "reduced" (time-independent) Schrödinger equation is elliptic, which will serve to remind you that my lectures have only covered a very limited area of the vast field of scientific computing. Nevertheless, I think that the fields I have mentioned are typical.

In reviewing the progress of the past 20–25 years in these fields, one becomes impressed with its slowness: *hardly a dent has been made in any of the problems listed above.* This fact may discourage some of you, but it does not depress me. On the contrary, I regard it as revealing infinite and inexhaustible opportunities for future discovery. Indeed, young men embarking on a career of research into scientific computing—even in the limited area of elliptic problems—need not fear that all problems will be solved in their lifetime. However, to make any contribution at all, they must accept the traditional standards of scientific research: a thoughtful and thorough appreciation of the previous contributions of others is a necessary prerequisite to publication of one's own ideas.

If these lectures have made more accessible to research workers previous contributions to the numerical solution of elliptic equations, they will have served their purpose.

REFERENCES FOR LECTURE 9

[1] J. H. ARGYRIS, *Energy Theorems and Structural Analysis*, Butterworths, London, 1960. (Reprinted from Aircraft Engineering, 1954–5.)
[2] G. BIRKHOFF AND E. H. ZARANTONELLO, *Jets, Wakes, and Cavities*, Academic Press, New York, 1957.
[3] E. CLEMENTI, *Ab initio computations in atoms and molecules*, IBM J. Res. Develop., 9 (1965), pp. 2–19.
[4] JULIAN D. COLE, *Perturbation Methods in Applied Mathematics*, Blaisdell, Waltham, Massachusetts, 1968.
[5] R. DAVIES, editor, *Cavitation in Real Liquids*, Elsevier, Amsterdam, 1964.
[6] H. EMMONS, *Critique of numerical modeling of fluid-mechanics phenomena*, Annual Rev. Fluid Mech., 2 (1970), pp. 15–36.
[7] D. GREENSPAN, editor, *Numerical Solution of Nonlinear Differential Equations*, John Wiley, New York, 1966.
[8] A. D. MCLEAN AND M. YOSHIMIME, *Computation of molecular properties and structure*, IBM J. Res. Develop., 12 (1968), pp. 206–233.
[9] H. SCHLICHTING, *Boundary-Layer Theory*, McGraw-Hill, New York, 1955.
[10] MILTON VAN DYKE, *Perturbation Methods in Fluid Mechanics*, Academic Press, New York, 1964.
[11] O. C. ZIENKIEWICZ AND Y. K. CHEUNG, *The Finite Element Method in Structural and Continuum Mechanics*, McGraw-Hill, New York, 1967.
[12] IEEE Trans. on Microwave Theory and Techniques, vol. 17, #8, Aug., 1969. (Special Issue on Computer-oriented Microwave Practices.)
[13] R. W. HOCKNEY, *The potential calculation and some applications*, Methods Comp. Phys., 9 (1970), pp. 135–211.
[14] E. CUTHILL, *Digital computers in nuclear reactor design*, Advances in Computers, 5 (1964), pp. 289–348.

QA
374
B57

JUL 17 1974